D0215623

WATER ON MARS

WATER ON MARS

MICHAEL H. CARR

U.S. Geological Survey
Menlo Park, California

New York   Oxford
Oxford University Press
1996

Oxford University Press

Oxford   New York
Athens   Auckland   Bangkok   Bombay
Calcutta   Cape Town   Dar es Salaam   Delhi
Florence   Hong Kong   Istanbul   Karachi
Kuala Lumpur   Madras   Madrid   Melbourne
Mexico City   Nairobi   Paris   Singapore
Taipei   Tokyo   Toronto

and associated companies in
Berlin   Ibadan

Copyright © 1996 by Oxford University Press, Inc.

Published by Oxford University Press, Inc.,
198 Madison Avenue, New York, New York 10016

Oxford is a registered trademark of Oxford University Press

All rights reserved. No part of this publication
may be reproduced, stored in a retrieval system, or transmitted,
in any form or by any means, electronic, mechanical,
photocopying, recording, or otherwise, without the prior
permission of Oxford University Press.

Library of Congress Cataloging-in-Publication Data
Carr, M. H. (Michael H.)
Water on Mars / Michael H. Carr
p. cm.
Includes bibliographical references and index.
ISBN 0-19-509938-9
1. Mars (Planet)—Geology.
2. Life on other planets.
3. Water.
I. Title.
QB641.C366   1996
551.48′0999′23—dc20   95-21541

1 3 5 7 9 8 6 4 2

Printed in the United States of America
on acid-free paper

One of the most intriguing aspects of Mars' evolution is the role played by water. The atmosphere contains only minute amounts of water, liquid water is unstable everywhere, and ice has been detected only at the north pole. Yet the surface shows abundant evidence of erosion by liquid water, and ground ice has been invoked to explain many additional features. The presence of water-worn landforms of different ages has led to speculation that Mars has undergone major changes in climate. Some researchers suggest that early Mars was warm and wet, and that conditions changed early to the frigid conditions of today. Others speculate that Mars has experienced large, episodic changes in climate throughout its history. Yet others are skeptical about any climate change. Similar disagreements prevail with respect to the amount of water present near the surface today and in the past. Estimates range from a low of 1 m spread evenly over the whole planet to well over a kilometer.* Uncertainties also exist about how the water inventory has evolved. Some maintain that Mars started out with a large inventory but lost most of it early by escape to space; others maintain that the abundant evidence of water erosion is a testament to the water's retention.

These issues have a profound importance for the formation of planets, the evolution of atmospheres, and the origin of life. It has been argued, for example, that as the Earth accreted, it accumulated an immense steam atmosphere but that most of this early steam atmosphere was lost to space, the remainder condensing to form the oceans. Other theories are that the oceans accumulated from cometary infall during the first half-billion years of Earth's history, or that oceans accumulated slowly with time. Better knowledge of how water evolved on Mars will better enable us to judge between the possibilities. Similar arguments apply to the atmosphere. Evidence that early Mars suffered a massive loss of atmospheric constituents because of high fluxes of extreme ultraviolet radiation from the Sun would imply that the Earth was similarly exposed.

Mars holds an additional special interest because of the possibility that life could have started there. The Viking missions to the surface of Mars in 1976 failed to detect any extant life or organic molecules despite sophisticated instruments with low detection limits. The high ultraviolet radiation, the oxidizing nature of the soil, and the lack of liquid water render the present-day surface very inhospitable. Hospitable conditions may, however, have prevailed in the past. Valley networks that resemble terrestrial valley systems suggest previous warm and wet conditions, particularly very early in the planet's history, at the time that life started here on Earth. Evidence of large floods implies the former presence of lakes, which could have harbored life. Thick stacks of sediments further imply that some lakes were long lived. Terrestrial hydrothermal systems commonly have rich biotas. Abundant evidence for sustained volcanism coupled with the strong possibility of pervasive ground ice argues that hydrothermal activity was common on Mars, and this appears to be confirmed by hydrothermal minerals in meteorites from Mars. Thus, the issue of whether life ever started on Mars is still very open, and the possibility that life did start depends crucially on the past availability of liquid water.

A book summarizing the history of water appears timely. Several new developments have occurred since previous summaries of the geology of Mars and its fluvial features (Carr, 1981,

_____

*For comparison, if the Earth's oceans were spread over the whole planet they would form a layer 3 km thick.

Baker, 1982) were published. Possibly the most important development has been recognition that a small group of meteorites (called SNCs and pronounced "snicks") come from Mars. The evidence for a martian origin will be elaborated upon later. But these meteorites have provided a wealth of new geochemical and mineralogic data pertinent to the history of Mars and the history of water in particular. Geologic interpretation of the surface features must be reconciled with these new data. In addition, the Russian Phobos mission has made new measurements of the surface and losses of gases to space from the upper atmosphere. Moreover, over the last decade the vast Viking data set has been examined in detail, thereby unearthing many new observations that must be explained. Modeling of different processes, such as climate change, circulation of groundwater, and planetary accretion, has also provided additional insights. Indeed, so much new information has been acquired and so much new analysis done that the possibilities seem to have grown substantially, our perceptions of the history of the planet have grown more complicated, and reconciliation of the observational data with evolutionary models is getting more difficult.

In this book I attempt to summarize the issues and highlight problems rather than solve them. The emphasis is on geology and geochemistry. Although a considerable amount of work has been done on the circulation of the atmosphere and seasonal changes in the distribution of water within the atmosphere, these topics are discussed only insofar as they may illuminate broader issues such as climate change and water inventories. The main concern here is with the past. How much water did the planet accrete, and how much was present near the surface at the end of accretion? How did the surface inventory change with time? Where is the water now? How and when did the different types of channels and valleys form? What climatic conditions were required for their formation? How and when did the climate change? The main records available to us to address these problems are the morphology of the surface, as revealed mainly in the Viking images, the chemistry of the atmosphere, and the chemistry and mineralogy of some near-surface rocks, the SNC meteorites.

Much of the discussion concerns photointerpretation and modeling, and usually neither technique leads to definitive results. In the absence of supporting geochemical and geophysical data, and information on lithology and the three-dimensional configurations of rock units, surface morphology is rarely susceptible to a unique and unambiguous interpretation, so one interpretation is rarely left unchallenged. Theoretical modeling of geologic processes may be even more suspect. The surface of a planet is the result of a succession of stochastic processes acting upon media, the near-surface materials, that are discontinuous at all scales from the microscopic to hundreds of kilometers. As a consequence, modeling of geologic processes is always overly idealistic, and judgments have to be made about the applicability of the results. Modeling of the atmosphere and deep interior, being less subject to problems of discontinuity and episodicity, is likely to be considerably more useful. Because of these problems and the general difficulty of understanding a remote, poorly explored object, the text is laced with qualifying words such as *may*, *suggests*, *possibly*, and *probably*. This may cause the reading to be tedious in places, but I want the reader to remain skeptical. There is very little here that is certain.

The book was written in the fall of 1994, and represents, as best as I can portray it, our state of knowledge at that time. I have of necessity, however, been somewhat selective. A recent book cited over 2,500 papers on Mars. I cannot claim to have read all these, let alone understand them, and I apologize for the many omissions. Different chapters were reviewed by Steve Clifford (Lunar and Planetary Institute), Ginny Gulick (NASA/Ames), Bruce Jakosky (University of Colorado), Jim Kasting (University of Pennsylvania), Chris McKay (NASA/Ames), and Ken Tanaka (USGS). Each added significantly to the accuracy of the final text. Sam Arriola (USGS) helped with the illustrations, and numerous coworkers provided originals of illustrations from papers that they had written.

Planetary science depends on getting missions approved and flown. Science comes only at the end when all the work has been done. Yet it sometimes seems that scientists have most of the fun and get most of the glory. I would like, therefore, to emphasize my appreciation to all those managers and engineers, who have worked so hard to get planetary missions approved, and to get spacecraft built, launched, and successfully flown. The science community owes them an enormous debt, for without their efforts we would still be left Earth-bound, wondering what those distant planets are really like.

Menlo Park, Calif.   M. H. C.
April, 1995

CONTENTS

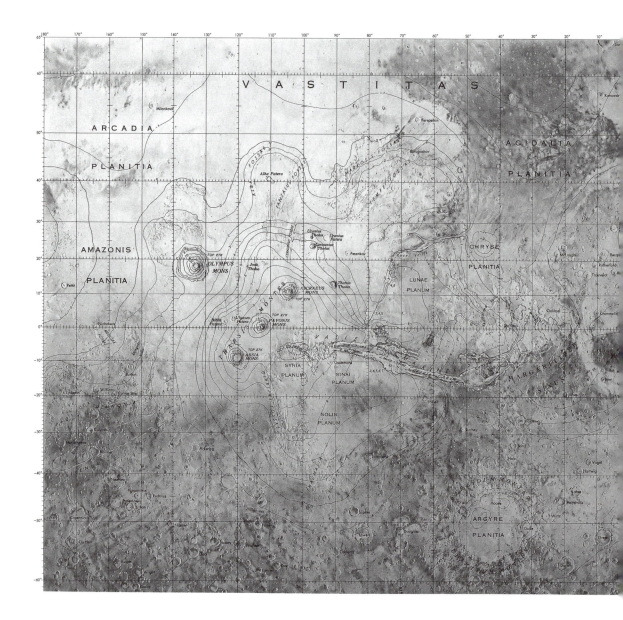

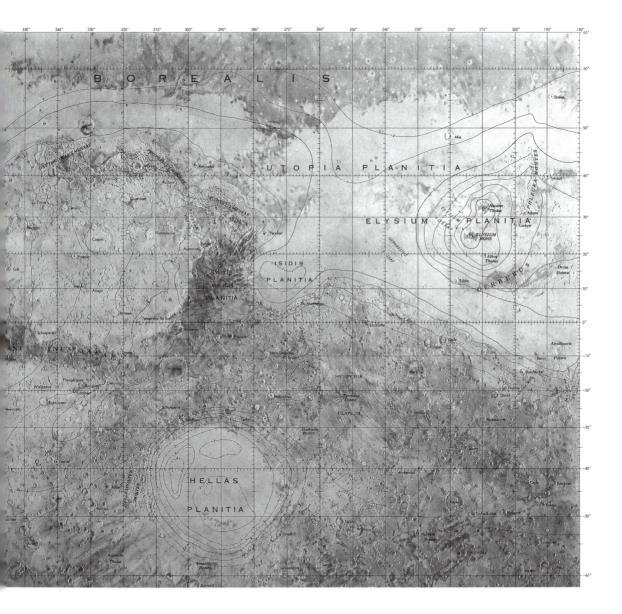

1 INTRODUCTION

This introduction provides an overview of current knowledge about the planet Mars. It is intended as a backdrop for the rest of the book, in which topics specifically related to water are discussed. Some major aspect of Mars' evolution, such as volcanism, tectonics, and atmospheric circulation, are summarized in a few paragraphs and then barely mentioned in the rest of the book. Other aspects, such as erosion of outflow channels and formation of the core, are picked up later because of their implications for the water story. For much more complete discussions of many of the topics introduced here, see Kieffer et al. (1992).

Historical Background

Since Galileo first observed Mars through a telescope early in the 17th century, water has been an essential part of our perception of Mars. By the late 18th century, as a result of systematic observations by many astronomers, notably Huygens, Cassini, and Herschel, the martian day had been accurately measured, the advance and retreat of the polar caps had been observed, and transient brightenings had correctly been attributed to dust and water clouds. It was recognized that the inclination of the axis was similar to the Earth's so that Mars must have seasons. In some respects, the perception of Mars was remarkably modern. Several prominent dark markings were, however, attributed to, and named as, seas or *maria*, and a dark collar around the polar cap that appeared in spring was thought to result from meltwater. Thus Mars was regarded as very Earthlike, a perception that persisted through the 19th century. Indeed, Mars was thought to be so Earthlike that several prominent scientists thought that intelligent beings might live there and speculated on ways to communicate with them.

A major shift in the perception of Mars came at the end of the 19th century. Several maps made at this time included linear features, termed *canali*. The most influential maps were those made by Schiaparelli during the 1877 opposition and published by Flammarion in 1892. They showed an elaborate global network of canals, which soon became the most renowned feature of the planet. The Boston astronomer, Percival Lowell, in a series of widely read books (Lowell, 1895, 1906, 1908) aggressively promoted the importance of the canals, drew ever more elaborate maps of them, and wrote of the possibility that the canals had been built by an advanced civilization in response to the progressive desiccation of their planet. Although the existence of the canals was challenged by many contemporary observers, the view that Mars had canals persisted. As recently as 1964 the astronomer Slipher (1964) published many photographs of Mars, taken at the observatory in Flagstaff, Arizona, established by Lowell, claiming that these photographs should remove all doubts about the canals. In 1971, at the time of the *Mariner 9* mission, canals were the most prominent features on some of the premission planning charts, on which we were to plot what we saw in the *Mariner 9* pictures. However, almost no traces of the canals were found in spacecraft images. They appear to have been largely the imaginings of observers, straining to perceive features at the resolution limit of their instruments (for an historical summary, see Kieffer et al., 1992).

The modern era of Mars exploration started on July 15th, 1965 when the *Mariner 4* spacecraft flew by Mars and took 22 fuzzy, low-contrast pictures of terrain that resembled the lunar highlands (Leighton et al., 1965). The results were disappointing to those expecting to see canals, oases, and evidence of flourishing vegetation. They seemed to indicate a cold, dry, dead planet of little geologic or biologic interest. These impressions seemed confirmed by the *Mariner 6* and *Mariner 7* missions, which took over 50 images, some with resolutions as high as 300 m/pixel (Collins, 1971; Leighton et al., 1969). Again, most of what was seen was a cratered surface, although there were some tantalizing glimpses of other kinds of terrain that were described as featureless or rubbly.

It was left to *Mariner 9* to reveal the enormous geologic diversity of the planet and to revive interest in its biologic potential. Originally, two U.S. spacecraft were to be sent to Mars in 1971. One was to be devoted to study of variable features, such as atmospheric activity and seasonal changes of the surface; the other was to systematically map the surface. The first spacecraft, *Mariner 8,* failed at launch; the second, *Mariner 9,* was successful beyond all expectations. When the spacecraft arrived at Mars in November 1971, almost the entire planet was engulfed in a dust storm. The only things clearly visible were the south pole and four dark spots in Tharsis. The latter, correctly interpreted as volcanoes poking up above the dust storm, were the first indications that Mars was not the geologically dead planet we had formerly thought. Fortunately, the dust started to clear early in January 1972, and systematic observation of the planet could start. Observations over the next several months revealed the geologically diverse planet that we know today, with its towering volcanoes, deep canyons, dry river beds, and extensive dune fields. It was evident that Mars had a complicated geologic history and that water had played a prominent role (for summary of pre-Viking missions, see Mutch, et al. 1976)

The Soviets sent two spacecraft to Mars in 1971, *Mars 2* and *Mars 3.* Each spacecraft carried a lander that it released to the surface just before being injected into Mars orbit. The first lander is thought to have crash-landed in the Hellas region. The second lander successfully soft-landed but failed after a 20-second transmission (Marov and Petrov, 1973). Plans for the orbiters to image the surface were frustrated by the global dust storm. The Soviets followed with four additional spacecraft in 1973. *Mars 4* failed to achieve Mars orbit. *Mars 5* was successful and functioned for 20 orbits, returning data of *Mariner 9* quality. *Mars 6* successfully landed but contact was lost just seconds before touchdown. The fourth spacecraft, *Mars 7,* flew by the planet (Snyder and Moroz, 1992).

The *Viking* mission was the next major step in the exploration of the planet. The main goal of the mission was to look for life. Two spacecraft were launched, each consisting of a lander and an orbiter. The landers were to go to the surface and perform a variety of experiments directed toward life detection. The orbiters were to observe the landing sites prior to landing to ensure that they were safe, to act as a relay station for lander data, and to observe the planet and landing sites so that the context in which the lander experiments were performed was understood. While the sophisticated lander experiments failed to detect life, the mission was enormously successful. All the lander experiments, except one seismometer, performed flawlessly, returning four years of data on the organic and inorganic chemistry of the soil, the response of the soil to various incubation conditions, the composition of the atmosphere, changes in the local scene, and the local meteorology. Meanwhile, the two orbiters systematically photographed the surface, mapped its thermal properties, and followed changes in the water content of the atmosphere. By the time that both orbiters had run out of attitude control gas in August 1980, they had taken over 50,000 pictures, mapped almost the entire planet at a resolution at least as good as 200 m/pixel, and sampled small fractions of the planet at resolutions as high as 9 m/pixel. The Viking science results are summarized in Carr (1981). For details of the planning and implementation of the Viking mission see Ezell and Ezell (1984).

In 1988 the Russians launched to Mars an elaborate spacecraft, designed primarily to study Phobos. It was hoped that the spacecraft would rendezvous to within 30–50 m of Phobos, make a wide range of remote-sensing observations, and release two landers to the Phobos surface to make direct measurements. After insertion into Mars orbit, the spacecraft successfully made observations of both Mars and Phobos, returning particularly useful data on the spectral and thermal properties of the martian surface, and on the erosion of the upper atmosphere by the solar wind. Unfortunately, contact was lost with the spacecraft during the complicated maneuvers to rendezvous with Phobos. The U.S. Mars Observer spacecraft, launched in 1992 met a similar fate. It was to have systematically mapped the gravity and magnetic fields; mapped the composition, spectral reflectivity, and elevation of the surface; and taken global and high-resolution pictures. Contact with the spacecraft was lost during the Mars orbit insertion maneuver.

Meanwhile, new information on Mars had come from an unexpected source. Nine anomalous meteorites, named SNCs for the locations (Shergotty, Nakhla, and Chassigny) where three of them were observed to fall, have properties in common and distinct from all other meteorites. They are all mafic volcanic rocks: Shergottites are basalts, Nakhlites are pyroxenites, and Chassigny is a dunite. They resemble each other in having similar mineralogies, similar Fe/(Fe + Mg) ratios, similar redox states, similar trace element patterns, and similar oxygen isotopic ratios. These properties are all distinctively different from other meteorites and terrestrial rocks. The property that was initially most suggestive of a martian origin was their age. They have crystallization ages of no more than 1.3 Gyr and possibly as young as 150 Myr. All other meteorites have 4.55 Gyr ages. A source had to be found that was volcanically active as recently as least 1.3 Gyr ago and Mars is the only likely candidate (for summaries, see Wood and Ashwal, 1981; McSween, 1985). The most convincing evidence for a martian origin, however, came later from the isotopic ratios of enclosed gases. Viking analyses show that the martian atmosphere has unique isotopic ratios for nitrogen and the various noble gases. Gases in the SNC meteorites have almost identical ratios after corrections for terrestrial contamination (Becker and Pepin, 1986; Bogard et al., 1984).

Astronomical Properties

Mars is the fourth planet from the Sun. In size it is intermediate between the Earth (6378 km radius) and the Moon (1738 km radius), approximating a triaxial ellipsoid with equatorial radii of 3394 and 3399 km, and a polar radius of 3376 km. The orbital period (year) is 687 Earth days, and the mean solar day is 24 hr 39.6 min. This results in a Mars year that has 669 Mars days or *sols*. The mean semi-major axis of the orbit is 1.524 A.U., so that the mean solar flux at Mars is 0.43 times that at the Earth.* For this flux the av-

erage surface temperature is 210 K, assuming a mean albedo of 0.25. For comparison, if the Earth had no atmosphere and the same albedo, it would have an average temperature of 259 K. The orbit is distinctly elliptical, having an eccentricity of 0.093 compared with the Earth's eccentricity of 0.017. At closest approach to the Sun (perihelion) the Mars-Sun distance is 1.381 A.U., or 206.5 million km; at its furthest distance from the Sun (aphelion), the Mars-Sun distance is 1.666 A.U. or 249.1 million km. As a result, the difference in solar flux at perihelion is 1.45 times that at aphelion.

The rotation axis is tilted 25° to the pole of the orbit plane, so that Mars, like the Earth, has seasons. Because of the eccentricity, the seasons vary significantly in length (Table 1-1). At present Mars is at perihelion at the end of southern spring, so that southern summers are shorter and hotter than those in the north. Time during the martian year is commonly referred to in terms of L_s, the areocentric longitude of the Sun. L_s is equivalent to the sun-centered angle between the line of the equinoxes and the position of Mars in its orbit, as measured from the vernal equinox (Figure 1-1). $L_s = 0$ is the start of northern spring, $L_s = 90°$ is the start of northern summer, and so forth.

Mars experiences long term variations in its rotation and orbital motions (see Chapter 6), and these may have significantly affected the stability and distribution of water at the surface. The rotational axis and the pole to the orbit plane both precess slowly, and the eccentricity changes in magnitude. Southern summers are presently shorter and hotter than those in the north, but the precessional cycle causes an alternation between the two hemispheres on a 51,000 year cycle. Thirty thousand years from now, northern summers will be the shorter and hotter. Changes in eccentricity affect the magnitude of the differences between the two hemispheres.

Variations in obliquity are particularly important for the history of water. Obliquity is the angle between the equatorial plane and the plane of the orbit. At low obliquities almost no solar insolation falls on the poles and they act as cold traps for volatiles, such as water. At high obliquities, the summer pole is tilted at a high angle toward the Sun and is constantly illuminated. It thus receives a large amounts of insolation,

---

*A.U. is an astronomical unit, defined as the mean distance between the Earth and the Sun, 149.5 million km.

Table 1-1. Mars Facts

Orbit semimajor axis	1.52366 AU
Eccentricity	0.0934
Obliquity	25.19°
Mean orbital period	686.98 Earth days
	669.60 Mars solar days (sols)
Mean solar day	24h 39.6m
Mass	6.4185×10^{23} kg
Mean radius	3389.92 km
Surface gravity at equator	3.711 m s^{-2}
Mean escape velocity	5.027 km s^{-1}
Total surface area	1.4441×10^{14} m^2
Ratio of total surface area to land area of Earth	0.976
Mean atmospheric pressure at surface	5.6 mbar or 560 Pa
Average mass of atmosphere	2.17×10^{16} kg
Mean atmospheric scale height	10.8 km
Solar constant at mean distance from sun	588.98 W m^{-2}

Mars seasons

Northern Spring	$L_s = 0\text{-}90^0$	199 days
Northern Summer	$L_s = 90\text{-}180^0$	183 days
Northern Fall	$L_s = 180\text{-}270^0$	147 days
Northern Winter	$L_s = 279\text{-}360^0$	158 days

Mars time-stratigraphic system.

Noachian-Hesperian boundary

Crater densities	200 > 5 km in diameter/10^6 km^2
	25 > 16 km in diameter/10^6 km^2
Absolute age	3.5 - 3.8 Gyr

Hesperian-Amazonian boundary

Crater densities	400 > 2 km in diameter/10^6 km^2
	67 > 5 km in diameter/10^6 km^2
Absolute age	1.8 - 3.5 Gyr

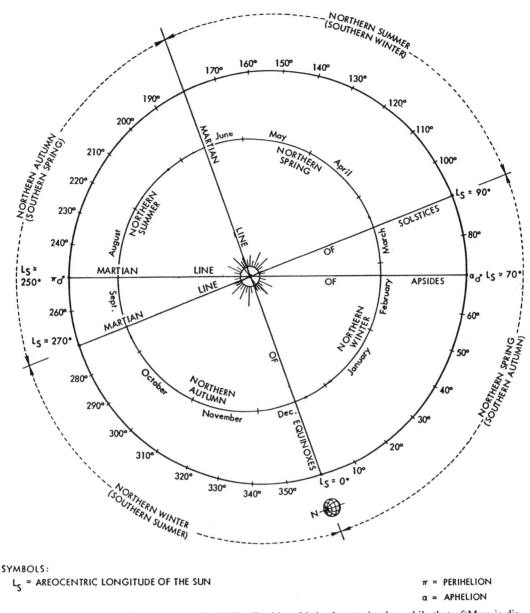

SYMBOLS:

L_S = AREOCENTRIC LONGITUDE OF THE SUN

π = PERIHELION

α = APHELION

Figure 1-1. The orbits of Mars and the Earth. The Earth's orbit is almost circular, while that of Mars is distinctly elliptical. Mars' seasons are referred to in terms of L_s, the areocentric longitude of the Sun, as shown. At present Mars is closest to the Sun at the end of southern spring, so southern summers are warmer and hotter than those in the north. (Adapted from Michaux and Newburn, 1972. NASA/JPL.)

which would tend to drive off any volatiles stored there. The present obliquity of Mars is 25.2°, but it may have experienced wide excursions in its obliquity, with values perhaps occasionally reaching as high as 60°.

Every 2.14 years Earth and Mars pass each other as they travel around the Sun. Because of the ellipticity of the orbit, the Earth-Mars dis-

tance at closest approach (opposition) varies by as much as a factor of two from opposition to opposition. The size of the Mars image as viewed from the Earth, therefore, varies correspondingly. The time interval between oppositions also affects Mars exploration, as minimum energy launch opportunities recur every 2.14 years. Launch opportunities occur, for example, in 1996,

1998, 2001, 2003, 2005, and so forth. Again, because of the eccentricity, the opportunities vary in terms of the energies required to deliver a specific payload to the planet.

The Atmosphere

The Mars atmosphere is thin. At the *Viking 1* landing site, which is at an elevation of 2 km below the Mars datum, the surface pressure ranges from 6.9 to 9 mbar, being at all times less than one hundredth of the average pressure at the surface of the Earth. The atmosphere is comprised of 95.3% CO_2, 2.7% N_2, 1.6% Ar, and only minor amounts of other constituents (Owen et al., 1977) (Table 1-2). Only minute amounts of water are present. The amounts range from less than 1 precipitable micron (pr μm) at the winter pole to 100 pr μm at the summer pole. The average is around 10 pr μm.[†] However, despite these small amounts, the atmosphere is close to saturation at all times. In winter, part of the CO_2 in the atmosphere condenses on the winter pole to form a seasonal CO_2 cap, thereby causing the variations in pressure that were observed at the two Viking landing sites (Figure 1-2). Because of the longer winter in the south, the southern winter CO_2 cap is more extensive than that in the north. At its maximum the northern seasonal cap extends to about 65°N, whereas at its maximum the southern seasonal cap extends to 50–60°S. As a result, more CO_2 is condensed out of the atmosphere during southern summer and surface pressures reach their minimum, 25% below their maximum value.

Figure 1-3 compares the globally averaged vertical structure of the Earth's atmosphere and Mars' atmosphere (Zurek, 1992). The lowest part of the Earth's atmosphere is called the *troposphere*. Temperatures within the troposphere decline with elevation and are controlled largely by radiative and conductive heat exchange with the surface and release of latent heat from the condensation of water vapor. In the *stratosphere*, above the *tropopause*, temperatures increase with elevation as a consequence of adsorption of ultraviolet radiation by ozone. Because of the re-

versed temperature gradient, there is little vertical mixing. Above the Earth's stratosphere, in the *mesosphere* the temperature gradient reverses again and temperatures decline with elevation, being controlled by radiative emission and absorption by CO_2. Finally, at the *mesopause*, the temperatures again start to increase with elevation, as heating is by conduction from above, where extreme ultraviolet radiation from the Sun is absorbed.

The vertical profile of the martian atmosphere is quite different, partly because of the lack of sufficient amounts of ozone to create a stable stratosphere and partly because of the effects of dust. When clear, temperatures in the lower part of the atmosphere decline with elevation, as in the Earth's troposphere. Up to an elevation of about 45 km, temperatures are controlled largely by exchange of heat with the ground. With the small amounts of water present, latent heating is negligible. From 45–110 km elevations, temperatures continue to fall but radiative emission and absorption by CO_2 dominate. Above the mesopause, at 110 km, the temperature gradient reverses, as on Earth, because the effects of absorption of extreme ultraviolet radiation high in the atmosphere become important. Just above the mesopause, at 125 km, is the *homopause*, above which atmospheric gases begin to separate diffusively. At still higher elevations is the *exosphere*, where the atmosphere is so thin that atoms and molecules are on ballistic trajectories and can escape. We will see later that diffusive separation of atmospheric components above the homopause and escape from the exosphere have resulted in a substantial enrichment of the atmosphere in deuterium and other heavy isotopes.

The conditions just described are for a clear atmosphere. The martian atmosphere holds at least a small fraction of dust at all times. The optical depth is typically 0.3–0.6, but may increase up to 5 during dust storms.[‡] Because the dust directly absorbs the Sun's radiation, heat transfer from the surface becomes less important in controlling the temperature and the vertical profile becomes

---

[†]A precipitable micron of water is such that if all the water in the atmosphere precipitated out it would form a layer 1 μm deep.

[‡]Optical depth is a measure of light absorbed by the atmosphere. For light entering the atmosphere with an intensity I_0 and an angle θ from the zenith, the intensity at the surface is $I = I_0 e^{-\tau/\cos\theta}$, where τ is the optical depth.

Table 1-2. Composition of the lower atmosphere
(Adapted from Owen, 1992)

Species	Abundance by volume
CO_2	0.9532
N_2	0.027
Ar	0.016
O_2	0.0013
CO	0.0007
H_2O	0.0003
Ar	5.3 ppm
Ne	2.5 ppm
Kr	0.3 ppm
Xe	0.08 ppm
O_3	0.04 to 0.2 ppm

Isotopic ratios	Mars	Earth
$^{12}C/^{13}C$	90 ± 5	(89)
$^{14}N/^{15}N$	170 ± 15	(272)
$^{16}O/^{18}O$ (atmospheric CO_2)	490 ± 25	(489)
$^{36}Ar/^{38}Ar$	5.5 ± 1.5	(5.3)
$^{40}Ar/^{36}Ar$	3000 ± 500	(296)
$^{129}Xe/^{132}Xe$	2.5	(0.97)
D/H	$(7.7 ± 0.3) \times 10^{-4}$	(1.5×10^{-4})

more isothermal. Diurnal temperature variations at the surface are also suppressed (Figure 1-4).

The circulation of the martian atmosphere has several components (Zurek et al., 1992). A north-south (meridional) flow results from seasonal exchange of CO_2 between the two poles, as 10–15% of the atmosphere condenses on the northern polar cap in northern winter and as much as 25% condenses on the southern polar cap in southern winter. Heating of the atmosphere at low to middle latitudes in the summer hemisphere causes air to rise there, promoting a seasonal meridional overturning (Hadley cell) that extends across the equator, thereby facilitating exchange of water vapor between the two hemispheres. The north-south flow and strong latitudinal thermal gradients in the midlatitudes of the winter hemisphere cause instabilities (ed-

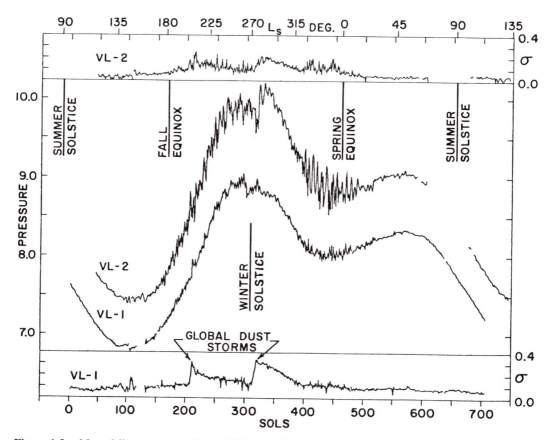

Figure 1-2. Mean daily pressures at the two Viking landing sites as a function of time of year. The broad variations are caused by condensation of CO_2 to form the seasonal caps. The annual pattern is dominated by the growth and dissipation of the southern cap, which is significantly larger than that in the north. The upper and lower panels show the standard deviations during a single day, and demonstrate the effect of enhanced diurnal heating and cooling when the atmosphere is laden with dust. (From Hess et al., 1980. Copyrighted by the American Geophysical Union.)

dies). As a result, eastward-propagating planetary waves develop and these are accompanied by strong westerly winds, high-altitude jet streams, and travelling storm systems. While the Viking landers were on the surface, the storms passed regularly on a roughly 3-day cycle. Modeling suggests that the northern storms are more intense than those in the southern midlatitudes. Other elements of the circulation include low-latitude, westward-propagating thermal tides, driven by the diurnal heating cycle, and quasi-stationary waves caused by the large-scale topography and large-scale variations in albedo.

At the Viking landing sites, outside the dust storm season, winds were typically a few meters per second, with daily maxima of 8–10 m s^{-1}. During the dust storm period, and at the times of the winter storms in the north, where the landing sites were, winds at the Viking sites were in excess of 10 m s^{-1} 10% of the time and gusts reached almost 40 m s^{-1}. The wind sensor was 1.6 m above the ground. At this elevation winds of 20-60 m s^{-1} are needed to cause saltation of surface grains, the exact value depending on the size of the grains and the surface roughness (Greeley et al., 1992).

When the atmosphere is clear, it is heated mostly from below, with the result that a convective boundary layer expands to a few kilometers thick during the day and then collapses at night. Most of the daily temperature variations damp out within 2 km of elevation above the surface. Because the atmosphere is so dry, latent heat effects are negligible and the vertical profile is close to the dry adiabat. The thermal stability of the base of the atmosphere at night, caused by the

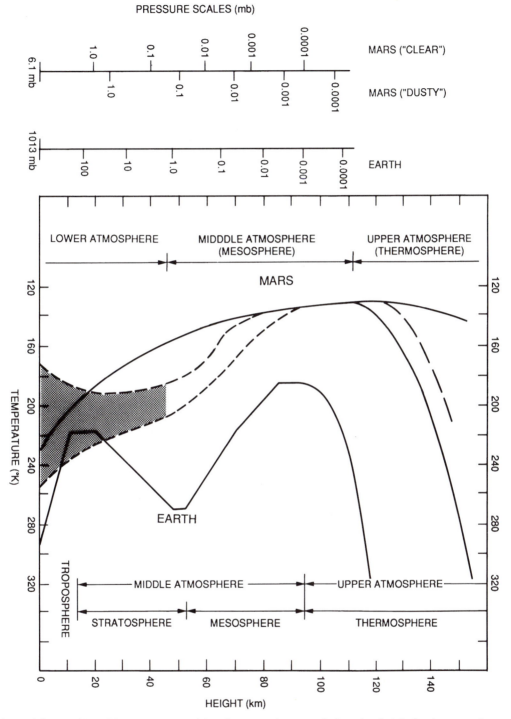

Figure 1-3. Profiles of temperature (x-axis) and pressure (upper scales) against height for the atmospheres of Earth and Mars. The solid line labeled Mars is for when the atmosphere is clear. The two branches in the upper martian atmosphere show the effects of the 11-year solar cycle. The dashed lines and shaded area indicate the range when the martian atmosphere is dusty. (From Zurek, 1992. Courtesy of R. Zurek, NASA/JPL.)

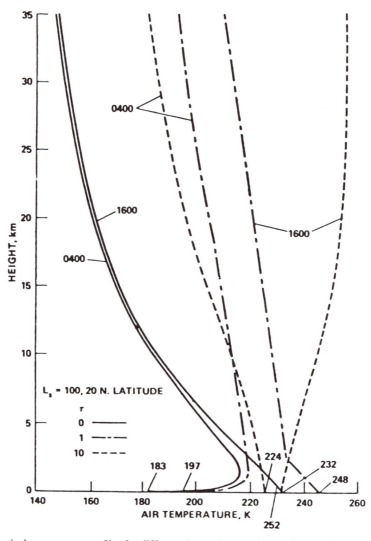

Figure 1-4. Vertical temperature profiles for different times of day and for different values of the optical depth, τ, at the start of the summer at 20° N latitude. When the atmosphere is clear, diurnal temperature variations are restricted to the lower few kilometers. When the atmosphere is dusty, the profiles become more isothermal and diurnal variations are significant through the entire profile. Figures at the base of the diagram are surface temperatures for the different profiles. (From Pollack et al., 1979. Copyrighted by the American Geophysical Union.)

extremely cold surface temperatures, decouples the atmosphere from surface friction and a strong nocturnal jet may develop, particularly at those times and at those places where the general circulation has a strong north-south component.

During southern spring and summer, dust storms tend to start at low latitudes, wherever there are large slopes and large gradients in surface albedo or thermal inertia. They also may start along the edge of the seasonal caps. Areas where dust storms have historically been initiated are the northwest rim of Hellas, the Claritas Fos-

sae region, and low-lying parts of Isidis Planitia (Gierasch, 1974). Dust storms may be local or they may grow to global proportions, as they did in 1971 at the start of the Mariner 9 mission and in 1977 during the Viking mission. Global dust storms can spread to encompass almost the whole planet in several weeks. The storms are most common in southern summer, close to perihelion, when summer temperatures are at their highest. During the 1976 dust storm the optical depth at the *Viking 1* landing site, far from the initiation site of the dust storm, rose from a value of 0.5 just before the storm to a high of 5 during the peak of the storm. The total amount of dust elevated into the atmosphere in the global storms is equivalent to a few micrometers spread over the whole planet (Kahn et al., 1992). At present, the storms appear to be effecting a net transfer of dust from the southern hemisphere to bright, low thermal inertia regions at low northern latitudes.

Surface Temperatures and Thermal Inertias

Because the atmosphere is dry, thin, and composed largely of CO_2, it has a low heat capacity and absorbs little of the Sun's incoming shortwave radiation or the outgoing long-wave radiation, at least when it is clear. As a result, for most of the year daytime surface temperatures are close to those expected from a simple balance between the solar radiation adsorbed at the surface and emitted infrared radiation. Daily mean temperatures at the equator are close to 215 K. Actual ground temperatures at the equator range mostly from about 160 K to 180 K at night to a peak temperature of about 260–280 K during the day. The exact temperatures depend on the latitude, the season, and the albedo and thermal inertia of the surface.

The radiometric albedo, normally designated A, is the fraction of the total incident solar radiation not adsorbed by the surface. Albedos of unfrosted ground range from 0.095 to 0.415, with preferred values at 0.135 and 0.275 (Kieffer et al., 1977). The thermal inertia (I) is a measure of the responsiveness of a material to changes in the thermal regime. It is defined as $(K\rho c)^{1/2}$, where K is the thermal conductivity, ρ is the density, and c is the specific heat of the material. Because the density and specific heat of rock materials do not vary greatly, most of the variations in thermal inertia are caused by variations in the thermal conductivity. Solid rocks have high thermal inertias; loose granular materials with abundant void spaces have low thermal inertia, conduction of heat through them being largely restricted to the contact points between grains.

Thermal inertias have historically been measured in units of calories (cal) $cm^{-2} s^{-1/2}$, but are normally referred to in inertia units for which the actual values have been multiplied by 10^3. One thermal inertia unit = 10^{-3} cal $cm^{-2} s^{-1/2}$ = 41.84 J $m^{-2} s^{-1/2}$. Thermal inertias of the martian surface range from 1 to 15 of these units (Kieffer et al., 1977; Palluconi and Kieffer, 1981). Bare rocks typically have thermal inertias in excess of 30, so all the martian surface is at least partly covered with loose materials. The thermal inertias of the martian surface cluster around two values, 6 and 2.5, corresponding, respectively, to the preferred albedo values of 0.135 and 0.275 (Kieffer et al., 1977). Figure 1-5 shows how surface temperatures change during the day as a function of latitude for the indicated values for albedo and thermal inertia.

Thermal inertias vary across the planet's surface (Figure 1-6). Two continental scale regions have low thermal inertias (fine-grained materials). These are the uplands of Arabia centered on 20°N, 330°W, and a broad region covering much of Tharsis, and the Amazonis and Arcadia Planitiae. Unusually high thermal inertias (coarse-grained materials) occur in the southern hemisphere within the Argyre and Hellas basins, and in the northern hemisphere in a broad region including Mare Acidalium, Chryse Planita, and the canyons, and in Utopia and western Elysium Planitia.

Lower thermal inertia surfaces heat up more quickly during the day and cool off more quickly at night. Thus a surface with a thermal inertia of 2.5 would have higher maximum and lower minimum temperatures than those in Figure 1-5, calculated for a thermal inertia of 6. But with a lower thermal inertia the temperature fluctuations at the surface damp out more quickly with depth, as is illustrated by Figure 1-7. Skin depth is a measure of how rapidly the temperature variations damp out at depth. The skin depth d, at which the temperature fluctuations have decreased to 1/e of their surface value, is defined as $(\kappa t/\pi)^{1/2}$, where κ is the diffusivity and t is the period of the temperature change. For a thermal inertia of 3 and 6, and typical values of ρ and c for soils (1.2 kg m^{-1} and 840 J kg^{-3}), the skin depths are, respectively, 6 and 10 cm for daily perturbations and 160 and 270 cm for annual perturbations. Thus the differences in surface materials implied by Figure 1-6 apply only to the upper few centimeters. They tell us nothing about the materials at greater depths.

At high latitudes in winter, surface temperatures are controlled mainly by condensation and sublimation of CO_2. In the fall, temperatures

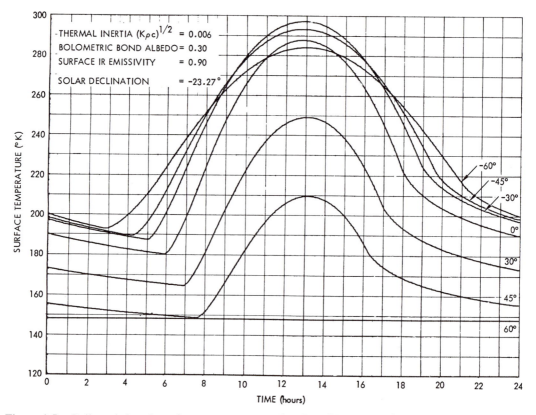

Figure 1-5. Daily variations in surface temperature as a function of latitude at $L^s = 270$, the start of southern summer. The surface albedo is 0.135 and a thermal inertia is 0.006 cal cm^{-2} s$^{-1/2}$, which are typical values for the martian surface.

drop until they reach 150 K, the frost point for CO_2, at which time CO_2 starts to condense onto the surface. The 150 K temperatures are maintained through winter as CO_2 continues to condense, and into spring until all the CO_2 has evaporated, at which time the temperatures rise rapidly. At the north pole in summer, evaporation of the CO_2 exposes a small water-ice cap. The temperature of the water-ice cap remains close to 200 K as water sublimes into the atmosphere. The maximum observed water content of 100 pr μm is over the northern summer pole. No residual water cap has been detected at the south pole. During the period of the Viking observations, the CO_2 cap in the south never completely evaporated so we do not know whether there is a small water-ice cap in the south.

Surface Topography

Mars has much larger variations in surface relief than does the Earth (Figure 1-8). Because there is

no sea level, elevations are referenced to a somewhat arbitrary datum, the level at which the average atmospheric pressure is 6.1 mbar, the triple point of water (Wu, 1978). The elevations currently available incorporate large errors. Temperature-pressure profiles can be derived from the decay of a spacecraft's radio signal as a spacecraft passes behind the planet when viewed from the Earth. These occultations provide the best measure of the position of the 6.1 mbar surface with respect tot he center of figure. Radar profiles provide very accurate relative elevations, but the profiles must be tied to the center of figure by some other means. Most of the gravity variations across the surface are caused by variations in topography, so gravity can be used to portray topography although only at a very coarse spatial resolution (Figure 1-8b). The highest spatial resolution topography (Figure 1-8a) is provided by numerous UV and IR spectroscopic measurements by *Mariner 9*. Elevations are derived on the assumption that a thicker atmosphere and

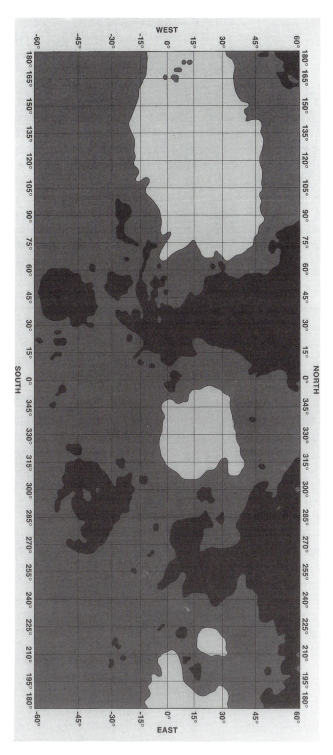

Figure 1-6. Generalized thermal inertia map. Surfaces within inertias of 0.001 to 0.004 cal cm^{-2} s$^{-1/2}$, suggestive of fine grained surface materials, occur in Arabia and in a large area including Tharsis and its environs. Relatively high thermal inertias of 0.008 to 0.015 cal cm^{-2} s$^{-1/2}$, suggestive of coarser materials are found within the canyons and the chaos areas to the east, within the low-lying northern plains, and in the impact basins, Hellas and Argyre. (Courtesy of F. Palluconi, NASA/JPL.)

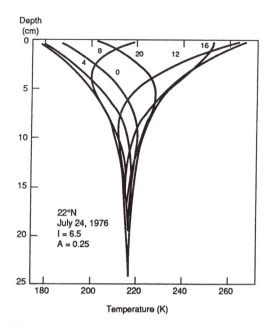

Figure 1-7. Temperatures below the surface for different times of day and the conditions shown. The temperatures rapidly converge at depth on the diurnal mean. The highest surface temperatures, reached just after noon (Figure 1-5), pertain only to the top several millimeters.

hence lower elevation would result in more scattering and adsorption by CO_2 in the atmosphere. These determinations are, however, prone to large errors because of the effects of dust and clouds. On the USGS map (Figure 1-8a) spectroscopic measurements are used to fill in gaps between the occultation measurements and radar profiles (Wu, 1978). Elevation errors are important for the water story in that interpretations have been made on the basis of slope reversals along stream profiles, slopes have been used to calculate flow discharges, and contours have been used to outline basins, and calculate water volumes. Some sense of the magnitude of the errors can be obtained by comparing Figures 1-8a and b.

The surface of Mars can be divided into two main components: (1) heavily cratered regions that have survived from the period when impact rates were very high, which by analogy with the Moon is thought to have ended 3.5–3.8 Gyr ago, and (2) more sparsely cratered regions that have been resurfaced since that time. The heavily cratered regions cover about two thirds of the planet's surface. This twofold division has been called the *global dichotomy*. Most of the heavily cratered regions are in the southern hemisphere and at elevations 2–5 km above the datum. Exceptions are the floors of the large basins of Hellas and Argyre, which extend to 5 km below the datum on the U.S.G.S. map. Many of the sparsely cratered plains are at elevations of 1–3 km below the datum. The most extensive of such plains are at high northern latitudes. Superimposed on this dichotomy are two broad bulges. The largest is the Tharsis bulge, a 10 km high, 6000-km wide rise centered on the equator at 110°W. The other is centered in Elysium at 30°N, 210°W. Large volcanoes in the Tharsis region reach elevations as high as 27 km above the datum. Thus, the elevation range is 32 km, as compared with 20 km for the solid surface of the Earth.

Geology

Geology is an interpretation, not an observation. Geologists attempt to reconstruct the history of a solid planet from the fragmentary record left at the surface. The main concerns are age and process, from which the sequence and nature of the events that led to the present configuration of the planet can be determined. With Mars the most extensive record we have is the morphology of the surface. Normally a geologist interprets the geology of a region armed with knowledge of the lithology, chemistry, mineralogy, and three-dimensional configuration of its rocks. Unfortunately, for most of Mars we have knowledge only of surface morphology and some properties of the upper surface, such as albedo and thermal inertia. While interpretation of surface morphology can be unambiguous, as in the case with the large volcanoes, it more often is not, so that many aspects of the geologic history are poorly understood. The following account focuses on topics not discussed in detail later. Features that have been attributed to the action of water or ice are mentioned only briefly. In particular, the channels, which are the main surface feature of interest with respect to water, are only noted in passing because they are treated more fully in subsequent chapters.

Relative age

One of a geologist's prime concerns is age. Although we have datable samples in the SNC me-

Figure 1-8. (a) The U.S.G.S. Global topographic map. Mariner-9 occultations and terrestrial radar profiles were combined with the gravity field, as determined by Mariner-9, to derive the large scale topography. Mariner-9 UV and IR spectroscopic measurements, from which the thickness of the atmosphere could be estimated, was used to interpolate between the occultation and radar points (Wu, 1978). Errors are estimated to be ±1 km at the equator and ±2 km at the poles, but could locally be significantly higher. Most of the southern cratered uplands are 3–5 km above the datum. The high-latitude northern plains are mostly 0–2 km below the datum. Superimposed on this dichotomy are a broad, 10 km high bulge centered on Tharsis and a much smaller, 5 km high bulge centered on Elysium. (b) The long wave length topography as inferred from the gravity field derived from Doppler tracking of the Mariner 9 and Viking 1 and 2 Orbiters (Smith and Zuber, 1994). The shape of the surface is provided by the gravity, the altitude is determined from occultations. The horizontal resolution is ~1500 km and the vertical resolution is ~1 km. Longitude is expressed in east-longitude. (Coutesy of D. Smith and M. Zuber, NASA/Goddard Space Flight Center.)

17

teorites, we do not know where on Mars they came from, so they are of little help in dating the features that we see. We are forced to extract age information from the morphology of the surface. Relative ages are determined from remote sensing in mainly two ways, from intersections and from the number of superimposed impact craters. Clearly, if a channel cuts a lava flow it is either the same age or younger than the flow. If a crater is superimposed on a surface, it is younger than that surface. But intersection relations are often ambiguous and other dating methods have to be used. Crater counting is the most commonly used way. Older surfaces have more superimposed impact craters. While this is a simple relationship, the method is, in practice, often difficult to apply. Because cratering rates have been very low since the end of heavy bombardment and because the cratering rate is strongly size dependent, large areas must be counted to get statistically significant results on images with the resolutions most commonly available. Ambiguities also result from partial or complete destruction of craters by erosion and burial, and as a result of other impacts. Such ambiguities multiply with the smaller, statistically more abundant craters because they are more easy to modify. Absolute ages from craters are very approximate.

A stratigraphic system for assigning ages to different parts of the martian surface has been devised that is calibrated against the numbers of superimposed impact craters (Tanaka, 1986). The relative ages so derived apply, of course, to the sculpting of the surface and not to the rocks themselves, although these are often identical, as with the formation of a lava flow. The geologic record has been divided into three time stratigraphic systems. Surfaces of Noachian age date back to the time of heavy bombardment. A 10^6 km^2 area of the surface that has survived from the top of the Noachian will have accumulated 200 impact craters with diameters larger than 5 km and 25 craters larger than 16 km (Table 1-1). A second system, the Hesperian system, refers to the oldest surfaces that postdate the end of heavy bombardment. A 10^6 km^2 area of the surface dating from the top of the Hesperian will have accumulated 67 impact craters with diameters larger than 5 km and 400 craters larger than 2 km. It will have accumulated too few 16 km diameter craters for meaningful counts. The youngest system is called the Amazonian system and obviously has lower numbers of craters than those just given for the top of the Hesperian. The three terms Noachian, Hesperian, and Amazonian are used throughout the book.

The absolute ages represented by these three systems are very poorly known, because we do not know how the cratering rates on Mars changed with time. The cratering history of the Moon is reasonably well known from crater counts and absolute dating of samples (see, for example, Wilhelms, 1987). Prior to 3.5 Gyr ago, the cratering rate was declining very rapidly (Figure 1-9), so much so that surfaces that date from 3.85 Gyr ago appear heavily cratered, almost saturated, whereas the maria from the Apollo 11 and Apollo 17 sites, with dates that range mostly between 3.6 and 3.75 Gyr ago, appear somewhat sparsely cratered. Low cratering rates have been maintained for the last 3.5 Gyr. The time before about 3.8 Gyr is referred to as the era of heavy bombardment. Unfortunately we do not know how to use the lunar curves to calibrate the martian crater counts. Mars is assumed to have experienced a similar cratering history to the Moon, with a high cratering rate tailing off rapidly around 3.8 billion years ago, but how rapidly the high cratering rate tailed off at the end of heavy bombardment is not known, nor is the cratering rate that has been maintained over the last 3 Gyr. Different cratering models give the age of the Hesperian-Amazonian boundary as 1.8–3.55 billion years and that of the Noachian-Hesperian boundary as 3.5–4.3 Gyr (Tanaka et al., 1992).

Highlands and plains

While Mars is much smaller than the Earth, the two planets have comparable land areas. Roughly two thirds of the surface, including almost all the southern hemisphere, is heavily cratered and has survived with only minor modification since the end of heavy bombardment. The cause of the hemispheric dichotomy into cratered uplands and sparsely cratered plains is unclear, but one plausible possibility is that the low areas of the north represent one or more scars from giant impact basins that formed at high northern latitudes at the end of accretion (Wilhelms and Squyres, 1984). The density of impact craters in the mar-

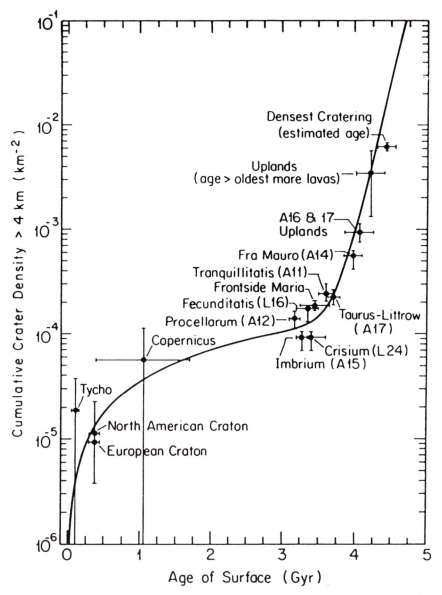

Figure 1-9. The cratering record of the Moon as derived from crater counts and absolute ages for the sample sites shown. The shape of the curve for Mars is probably similar but the scale on the vertical axis is unknown. (Reprinted with permission from Chyba, 1991. Copyright 1991 by the American Association for the Advancement of Science.)

tian highlands is comparable to the lunar highlands. The surface clearly dates back to the very earliest history of the planet, prior to 3.8 billion years ago, when impact rates were high.

The martian highlands differ from the lunar highlands in several ways. First, sparsely cratered plains are more common between the larger craters in the martian highlands. Some of the intercrater plains may be volcanic, as indicated by the occasional flow front and wrinkle ridge, which resemble those on the lunar maria. But clear evidence of volcanic plains is rare. Most of the intercrater plains are likely to be a mixture of impact ejecta and materials that have accumulated between the craters as a result of mass wasting, and fluvial and eolian processes. Other differences between the lunar and martian highlands concern the craters. The ejecta around lunar craters generally has a coarse, hummocky texture at the rim crest and grades outwards into a finer, randomly hummocky or radial texture. On Mars, the ejecta around craters 2–50 km in diameter is commonly arrayed in discrete lobes, each lobe being outlined by a low ridge. This is true of almost all well-preserved martian impact craters in this size range, irrespective of location.

Two reasons have been suggested for the characteristic martian ejecta patterns. The first is that impact craters above a certain size penetrate the permafrost zone and eject water-laden or ice-laden materials that tend to flow across the surface following ejection from the crater (Carr et al., 1977). This will be amplified upon later. The second suggestion, based on wind-tunnel experiments, is that interaction between the ejecta and the atmosphere causes the flowlike patterns (Schultz and Gault, 1984). Another difference between craters in the lunar and martian uplands is their state of preservation. Most of the craters in the martian uplands are highly degraded. This has been taken as one indication of high erosion rates and climatic conditions very different from the present during the early epoch represented by the cratered uplands. Yet another difference between the lunar and martian highlands is the presence within the martian highlands of numerous branching valley networks. At low latitudes they are almost everywhere within the highlands. They superficially resemble terrestrial river valleys and suggest the former presence of running water. They will be discussed at length in subsequent chapters.

Most of the extensive plains, as opposed to the intercrater plains in the uplands, are in the northern hemisphere. The number of superimposed craters on these plains varies substantially, indicating that the plains continued to form throughout much the history of the planet. The plains are diverse in origin. The most unambiguous in origin are those on which numerous flow fronts are visible. They are clearly formed from lava flows superimposed one on another, and are most common around the volcanic centers of Tharsis and Elysium. On other plains, such as the Lunae Planum and Hesperia Planum, flows are rare but wrinkle ridges like those on the lunar maria are common. These are also assumed to be volcanic, although evidence for a volcanic origin is weak. Wrinkle ridges are evidence of deformation not

volcanism. The vast majority of the low-lying northern plains lack obvious volcanic features. Instead they are curiously textured and fractured. Many of their characteristics have been attributed to the action of ground ice or to their location at the ends of large flood features, where large lakes must have formed after the floods and sediments must have been deposited. In some areas, particularly around the north pole, dune fields are visible. In yet other areas are features that have been attributed to the interaction of volcanism and ground ice. Thus the plains appear to be complex in origin, having been variously formed by volcanism and different forms of sedimentation, and then subsequently having been modified by tectonism and by wind, water, and ice.

Volcanism

The two most prominent volcanic provinces are Tharsis and Elysium. Three large volcanoes are close to the summit of the Tharsis bulge, and Olympus Mons, the tallest volcano on the planet, is on the northwest flank. All these volcanoes are enormous by terrestrial standards. Olympus Mons (Figure 1-10) is 550 km across and 27 km high, and the three others have comparable dimensions. Lava flows and lava channels are clearly visible on their flanks, and each volcano has a large, complex summit caldera. They all appear to have formed by eruption of fluid lava with little pyroclastic activity. The large sizes of the volcanoes have been attributed to the lack of plate tectonics on Mars (Carr, 1973) and to the buoyancy of the martian magmas with respect to edifices they build (Wilson and Head, 1994). The small number of superimposed impact craters on the flanks of the large Tharsis volcanoes indicates that their surfaces are relatively young. Although their surfaces may be young, the large Tharsis volcanoes have probably been growing throughout much of Mars' history, as indicated by the number of impact craters on flows peripheral to the volcanoes but derived from them (Carr et al., 1977b). Several smaller volcanoes occur throughout Tharsis, in addition to the largest ones just described.

To the north of Tharsis is Alba Patera, the largest volcano on the planet in areal extent. It is roughly 1500 km across but may be only a few kilometers high. Flows are visible on parts of its

Figure 1-10. Olympus Mons, the largest volcano on Mars, is 27 km high and 550 km across. *~ 87,000 feet* The scarcity of impact craters on the surface indicates that the surface is young (Amazonian). The presence of this and other volcanoes indicates that the interior of Mars has continued to outgas throughout the planet's history, up to the present (Viking Orbiter frame 646A28).

flank, but elsewhere the volcano flanks are dissected by numerous branching valleys. These valleys, resembling terrestrial fluvial valleys, are relatively young and have been used to argue for warm and wet conditions late in martian history. The easily eroded, channeled deposits have been interpreted as ash (Wilson and Mouginis-Mark, 1987). Densely dissected deposits on other volcanoes such as Ceraunius Tholus in Tharsis,

Hecates Tholus in Elysium, and Tyrrhena Patera in the southern highlands, have also been interpreted as ash. Thus Mars appears to have experienced both the Hawaiian style of volcanism, involving mostly quiet effusion of fluid lava, and more violent, pyroclastic eruptions, which result in deposition of extensive ash deposits.

Elysium Mons, the largest volcano in the Elysium province, appears to be a shield volcano formed largely of fluid lava. However, huge channels start at the periphery of the volcano and extend northwest, down the regional slope for hundreds of kilometers. The channels have streamlined forms and enclose teardrop-shaped islands. Similar large channels start adjacent to Hadriaca Patera on the rim of the large impact basin, Hellas, and extend for hundreds of kilometers down into the Hellas basin. All these channels have the characteristics of large floods. They are thought to have formed by massive release of water following melting of ground ice by the volcanoes. Numerous other features in the Elysium area and elsewhere have also been interpreted as a result of volcano-ice interactions (Squyres et al., 1987).

The rates of volcanism on Mars are much lower than those on Earth. Greeley (1987) and Greeley and Schneid (1991) estimated that roughly 6×10^7 km^3 of lava have accumulated on the martian surface since the end of heavy bombardment and that the average extrusion rate was 0.016 km^3 yr^{-1}. Taking into account intrusive rocks, the average magma production rate is likely to be about 10 times this figure. For comparison, the Earth has produced roughly 30 km^3 yr^{-1} for the last 180 Myr (Sclater et al., 1980). The rate of volcanism on Mars per unit mass of the planet is about 0.05 times that on the Earth.

Tectonism

While plate tectonics have been invoked to explain the northern lowlands (Sleep, 1994), most of Mars lacks obvious manifestations of plate tectonics such as linear mountain chains, subduction zones, large transcurrent faults, and an interconnected system of ridges. The most widespread indicators of surface deformation are normal faults, indicating extension, and wrinkle ridges, indicating compression. The most obvious deformational features are those associated with the Tharsis bulge. Around the bulge is a vast system of radial graben that affects about a third of the planet's surface. The graben are particularly prominent north of Tharsis, where many are diverted around the volcano Alba Patera to form a fracture ring. Circumferential wrinkle ridges are also present in places, particularly on the east side of the bulge in Lunae Planum. Both the fractures and the compressional ridges are believed to be the result of stresses in the lithosphere caused by the presence of the Tharsis bulge. No comparably extensive system of deformational features occur around Elysium, but fractures occur in other places where the crust has been differentially loaded, such as around large impact basins, such as Hellas and Isidis, or around large volcanoes, such as Elysium Mons and Pavonis Mons.

The vast canyons on the eastern flanks of the Tharsis bulge are the most spectacular result of crustal deformation. The canyons extend from the summit of the Tharsis bulge eastward for 4000 km until they merge with chaotic terrain and large channels south of the Chryse basin. In the central section, where several canyons merge, they form a depression 600 km across and several kilometers deep. Although the origin of the canyons is poorly understood, faulting clearly played a major role. The canyons are aligned along the Tharsis radial faults, and many of the canyon walls are straight cliffs, or have triangular faceted spurs, clearly indicating faulting. Other processes were also involved in shaping the canyons (Figure 1-11). Parts of the walls have collapsed in huge landslides; other sections of the walls are deeply gullied. Fluvial sculpture is particularly common in the eastern sections. Faulting may have created most of the initial relief, which then enabled other processes, such as mass wasting and fluvial action, to occur. Creation of massive fault scarps may also have exposed aquifers in the canyon walls and allowed groundwater to leak into the canyons, thereby creating temporary lakes, as evidenced by thick sequences of layered sediments.

Poles

The poles are distinctively different from the rest of the planet. At each pole and extending outward to about the 80° latitude circle is a thick stack of layered sediments. They are estimated to be at least 1–2 km thick in the north and at least 4–6 km thick in the south (Dzurisin and Blasius, 1975). Incised into the smooth upper surface of the deposits are numerous valleys and low escarpments. These curl out from the pole in a counterclockwise direction in the north and a predominantly clockwise direction in the south.

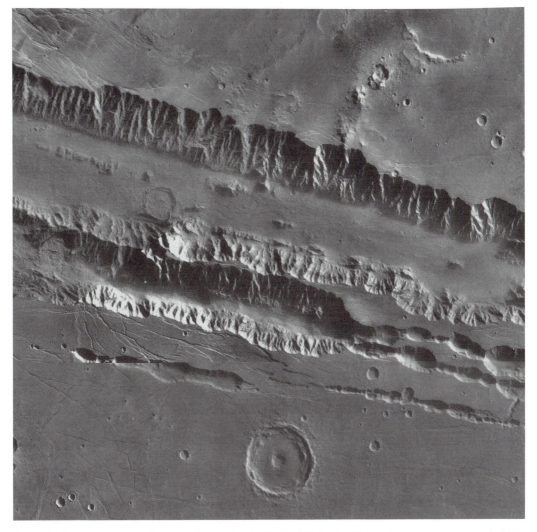

Figure 1-11. A section of Coprates Chasma in the Valles Marineris. The main canyon seen here is 90 km across and 6 km deep. The canyons are thought to have formed largely by faulting, and a fault scarp is clearly visible at the base of the far wall. The precise mode of formation is, however, not known. Particularly puzzling are the lines of closed depressions at the bottom of the picture and how these seemingly evolve into true canyons. The canyon walls are deeply gullied but no erosional debris can be seen on the canyon floor. Lakes may have formed upstream of this section and catastrophically drained through this part of the canyon (see Chapter 3). Faint sinuous markings can be seen on the floor of the main canyon.

Between the valleys, which are roughly equally spaced and 50 km apart, the surface of the deposits is very smooth and almost crater free. For most of the year the layered deposits are covered with CO_2 frost, but in summer they become partly defrosted. The layering is best seen as a fine horizontal banding on defrosted slopes. The deposits are believed to be composed of dust and ice, with the layering caused by different proportions of the two components. The scarcity of im-

pact craters indicates a relatively young age, although the age of the deposits could still be on the order of hundreds of millions of years (Plaut et al., 1988).

The cyclic sedimentation, implied by the layering, suggests that the origin of the deposits may be connected in some way with the changes in obliquity referred to earlier. At higher obliquities, much more insolation falls on the poles than at the lower obliquities. The capacity of the high-latitude regolith to hold adsorbed CO_2, the size of the CO_2 cap, and the atmospheric pressure probably all change with obliquity. These changes could affect global wind regimes, dust storm activity, and sedimentation rates at the poles, thereby causing cyclic sedimentation. Layering would also have been caused by events such as a floods, volcanic eruptions, and cometary impacts that introduced large amounts of water vapor into the atmosphere, because they would all have resulted in deposition of ice-rich layers at the poles.

Views from the surface

At the time of writing we had close-up views of the surface only at the two Viking landing sites. The *Viking 1* lander is located at 22.5°N, 48°W, in a sparsely cratered area with wrinkle ridges, which from the orbiter looks like a lunar mare (Binder et al., 1977). The surface has a rolling topography and is strewn with rocks in the centimeter to meter size range. They have spectral characteristics consistent with basalts. Between the rocks is mostly fine-grained material, which is also present as drifts. The surface of the drift material, is partly cemented to form a duricrust, indicating that the more volatile and soluble components have migrated within the soil profile. Several small areas, seemingly free of the drift material appear to be bedrock exposures. Although volcanic, the blocks have probably been ejected from impact craters. They are more common than on lunar maria because the atmosphere protects the surface from the erosive effects of micrometeorite bombardment. The second Viking lander is located at 48°N, 225.6°W, seemingly on a debris flow from the large impact crater Mie, 170 km to the east (Mutch et al., 1977). The view from the lander is of a flat plain, which is almost featureless except for numerous rocks, mostly in the 10–20 cm size range, distributed uniformly over the entire scene.

The poorly consolidated materials at the two landing sites are remarkably similar in composition (Table 1-3), which suggests that the material has been homogenized over the whole planet by repeated participation in dust storms over much of the planet's history. The Viking landers did not carry an experiment to determine mineralogy. It must be inferred indirectly from the chemical analyses, from the sparse spectral information available, from the behavior of the results of the Viking GCMS (Gas Chromatograph Mass Spectrometer) and biology experiments, and from weathering products found in SNC meteorites. Palagonite and montmorillonitic clays are the materials that have been most commonly proposed as the main constituents of the fine-grained material. Palagonite is a poorly crystalline or cryptocrystalline alteration product of basalts. Minerals that have been identified in terrestrial palagonites include montmorillonitic clays, such as nontronite, iron oxyhydroxides, serpentine, plagioclase, and calcite. Basaltic glass is also commonly present. The poorly defined, poorly crystallized assemblage appears to be the product of low-temperature (25–150°C) hydration and selective dissolution of the parent basaltic glass. Montmorillonitic clays are a major component of palagonite. The alternate clay model differs from the palagonite model largely in the degree of crystallinity that is visualized. Iron is a major component of the soil and the principal cation in the clays. But iron is probably also present, mostly as Fe^{3+}, in other phases, such as amorphous fine-grained oxides and oxyhydroxides, and iron minerals such as ferrihydrite, goethite, and hematite. The presence of large amounts of Cl and S in the soils at the two Viking landing sites indicates the presence of soluble salts, specifically chlorides and sulfates. The analyses are permissive of small amounts of carbonates and nitrates, but the landers had no means of detecting them.

SNC Meteorites

Nine of the ten known SNC meteorites are volcanic rocks (McSween, 1985, 1994) with radiometric ages that range from 0.15 to 1.3 Gyr (Table 1-4). A recently recognized tenth SNC meteorite (ALH84001) is an orthopyroxenite with a radiometric age of 4.5 Gyr (Nyquist et al., 1995). Shergottites are named for several stones that fell in Shergotty, India in 1865, but two additional shergottites have been recovered from Antarctica. Three shergottites are medium-grained basalts or diabases, consisting mostly of clinopyroxene (pigeonite and augite) and maskelynite, a shock-metamorphosed plagioclase.

Table 1-3. Chemical composition of Martian Soil.

(Adapted from Banin et al., 1992)

Constituent	Estimated Concentration (%)	Source
SiO	43.0	Direct soil analysis by Viking XRFS
Al_2O_3	7.2	"
Fe_2O_3	18.0	"
MgO	6.0	"
CaO	5.8	"
TiO_2	0.6	"
SO_3	7.2	"
Cl	0.6	"
K_2O	0.2	Analysis of Shergotty meteorite
P_2O_5	0.8	"
MnO	0.5	"
Na_2O	1.3	"
Cr_2O_3	0.2	"
Sum of above	91.4	
CO_3	< 2	Estimated from Viking Labelled Release
H_2O	0-1	Estimated from Viking GCMS
NO_3	?	

Olivine is a significant phase in only one shergottite, EETA79001. Minor amounts of Fe-Ti oxides, chromites, sulfates, and phosphates are also present. A hydrous amphibole kiersutite, found in melt inclusions trapped within the pyroxenes, is important for the water story because the water is enriched in deuterium, like the water in the atmosphere (Watson et al., 1994). Two additional shergottites are lherzolites, which have medium-grained olivine, and chromite enclosed in large orthopyroxenes. Nakhlites are named for the shower of about 40 stones that fell in El Nakhla, Egypt in 1911. The original stones have been supplemented by two additional finds (Lafayette and Governador Valadares). They are pyroxenites with minor amounts of olivine. Interstitial minerals are similar to those in the shergottites. The one Chassigny known fell in France in 1815. It is a dunite, consisting mostly of olivine with minor amounts of orthopyroxenes and clinopyroxenes. The additional Antarctic meteorite, ALH84001, that was recently recognized as belonging to the SNC meteorites, is a brecciated coarse-grained pyroxenite (Mittlefehldt, 1994). All the SNC meteorites have cumulate textures, that is, they formed as a result of concentration of crystals derived by fractional crystallization from a melt. Postcumulus minerals are found in the interstices

Table 1-4. The SNC Meteorites.

(Adapted from Gooding, 1992)

Meteorite	Date of fall/find	Mass Recovered (kg)	Radiometric Age (10^9 yrs)	Cosmic-ray Exposure age (10^6 yrs)	Petrology
Shergotty	Fell, 1865, India	5	165–205; 350	1.6–2.8	Shock-metamorphosed basalt
Zagami	Fell, 1962, Nigeria	23	116–230	1.9–3.1	Similar to Snergotty
ALHA77005	Found, 1977, Antarctica	0.48	187	1.9–3.0	Similar to Sherg. but w/ dark/light lith., cum. mins.
EETA79001	Found, 1979, Antarctica	7.9	150–185	0.35–0.99; 2.9	Similar to Sherg. but w/ igneous contact, melt, cum. mins.
Nakhla	Fell, 1911, Egypt	40	1240–1370	10.7–11.9	Pyroxene-olivine cumulate or coarse lava.
Lafayette	Found, 1931, Indiana	0.6	1330	10.5–11.9	Similar to Nakhla
Gov. Valadares	Found, 1958, Brazil	0.16	1320	6.3–9.7	Similar to Nakhla
Chassigny	Fell, 1815, France	4	1230–1390	9.6–11.8	Olivine cumulate

Sherg. = Shergotty; lith. = lithology; cu. mins. = cumulative minerals.

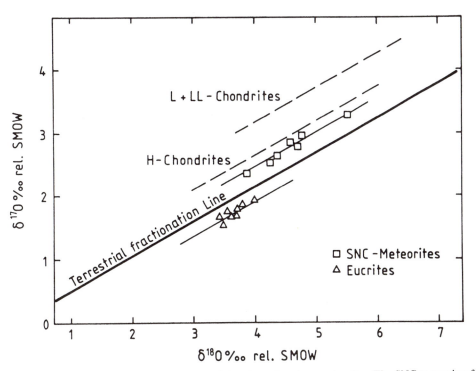

Figure 1-12. Oxygen isotope variations in terrestrial rocks and various meteorites. The SNC meteorites form a distinct coherent class with isotopic patterns different from other meteorites or terrestrial rocks. (From Clayton and Mayeda, 1983. Reproduced with permission from Elsevier Science.)

between the larger crystals. The modest crystal grain sizes indicate that the SNC parent magmas were emplaced at shallow depths within the crust or extruded onto the surface as phenocryst-rich lavas.

The case for a martian origin of these rocks was mentioned briefly earlier in the chapter: (1) Several of them were observed to fall and thus are known to be meteorites. (2) They are basaltic rocks with ages ranging mostly from 150 Myr to 1.3 Gyr and thus had to originate on a body that was volcanically active during this time. (3) They have oxygen isotope patterns that are distinctively different from those of terrestrial rocks and other meteorites (Figure 1-12). (4) They have element abundance patterns distinctively different from the Earth, being more enriched in moderately volatile elements and more depleted in chalcophile elements than typical terrestrial volcanic rocks (Wänke, 1981). (5) Gases enclosed in these meteorites have isotopic and chemical compositions (Bogard and Johnson, 1983; Bogard et al., 1984; Becker and Pepin, 1984) almost identical to those in the martian atmosphere (Figure 1-13) as measured by Viking (Owen et al., 1977). This evidence is almost absolutely conclusive of a martian origin, for not only must the source be a body that was volcanically active in the last 1.3 Gyr, but it must also have an atmosphere almost identical to that of Mars, which implies that the body had to have a similar evolutionary history as Mars. The $^{36}Ar/^{40}Ar$ ratio, for example, is the result of partial retention of early Ar, the potassium content of the interior, and the outgassing history, all of which would have to be similar to Mars. We can only conclude that the SNC meteorites came from Mars.

The SNC meteorites are believed to have been ejected from Mars during one or more large impact events. The ejection velocities were such as to allow them to escape from Mars into Earth-crossing heliocentric orbits from which they were ultimately captured by the Earth. The low levels of shock experienced by the meteorites indicates that they were ejected not from close to the point of impact but as spallation fragments from around the periphery of the impact site (Melosh, 1984). Exposure ages of the meteorites

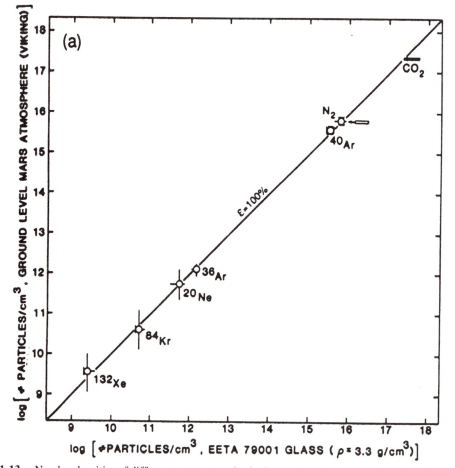

Figure 1-13.   Number densities of different gaseous species in the martian atmosphere plotted against the number densities in gases extracted from the SNC meteorite EETA 79001. The straight line relationship over many orders of magnitude indicates similar compositions in the two samples and represents compelling evidence that this meteorite came from Mars. (From Wiens and Pepin, 1988. Reproduced with permission from Elsevier Science.)

fall into three groups of 0.5, 2.6, and 11 million years. One possibility is that the SNCs were ejected from Mars during a large impact 11 million years ago and that some of the ejecta broke up as a result of collisions with other objects in space 0.5 and 2.6 million years ago (McSween, 1985).

Interior Structure and Heat Flow

The dimensions of the martian core are constrained by the planet's mean density (3933 kg m^{-3}) and its dimensionless moment of inertia (0.365). The constraints are satisfied on one extreme by a small (1300 km radius) metallic iron core, with a density of 8090 kg m^{-3} and consti-

tuting 14.8% of the planet's mass, or, on the other extreme, by a large (2000 km radius) iron sulfide core, with a density of 5770 kg m^{-3} and constituting 26.3% of the planet's mass. Evidence from SNC meteorites shows that the martian mantle is depleted in siderophile and chalcophile elements, which preferentially enter metallic and sulfide melts, respectively. The depletion in chalcophiles suggests a core size closer to the larger values permitted by the moment of inertia than to the smaller values (Treiman al., 1986).

   U-Pb data and Rb-Sr data on SNC meteorites (Shih et al., 1982; Jagoutz, 1991) indicate that core formation and global differentiation took place 4.4–4.6 Gyr ago, essentially at the end of accretion. Core formation would have released

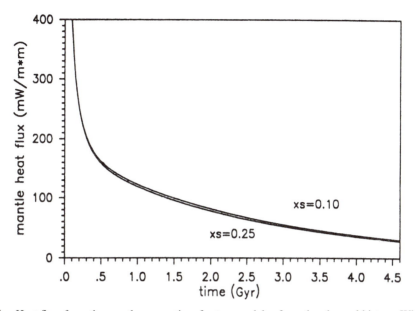

Figure 1-14. Heat flow from the mantle versus time for two models of martian thermal history. When the geologic record emerged, around 3.8 Gyr ago, the heat flow was five times that of today. (From Schubert et al., 1992. Reproduced by permission from the University of Arizona Press.)

sufficient heat to raise the temperature of the planet by 300 K (Solomon, 1979). Global differentiation implies that Mars was hot at the end of accretion, and modeling of cooling histories suggests that during its first few 100 Myr the planet was cooling rapidly as a result of a hot and vigorously convecting mantle (Stevenson et al., 1983). The heat flow fell from several hundred milliwatts per square meter to around 150 mw m^{-2} at the end of the heavy bombardment, since which time it has fallen slowly to its present value of around 30 mw m^{-2} (Figure 1-14). The lithosphere probably grew rapidly from close to zero at the end of accretion, to a few tens of kilometers at the end of heavy bombardment, and to the 100–200 km estimated for the thickness at present.

2 THE PRESENT WATER CYCLE AND STABILITY RELATIONS

This chapter describes the present-day water cycle on Mars, insofar as we know it, and explores some implications of present day conditions for the stability of water and water-ice near the surface. The present water cycle has been inferred largely from observations of the amount of water vapor in the atmosphere. This may represent only a minute fraction of that near the surface, but it is the only part of the inventory that we can measure directly. Though small, the water content of the atmosphere is important because it provides clues about the stability of water (ice) in the near-surface materials and so has long-term geologic implications and implications for future Mars exploration. The chapter also examines the holding capacity of the surface materials for water and ice, and how water might migrate through these materials, both on a short time scale and over geologically significant time scales.

Circulation of Water in the Atmosphere

Much of what we know of the circulation of water in the martian atmosphere is derived from the observations of the Mars Atmospheric Water Detection experiments (MAWD) on the Viking Orbiters (Jakosky and Farmer, 1982). The instruments measured the amount of water vapor in the atmosphere along their line of site by comparing the absorption of solar radiation in the 1.4 μm band with theoretically predicted absorptions for different amounts of water in the atmosphere. The line-of-site observations were translated into vertical column abundances from the known viewing geometry. In a low-resolution mode, the data were binned in 10° boxes (approximately 600 × 600 km); in a high-resolution mode they were binned in $2\frac{1}{2}°$ boxes (approximately 150 × 150 km). The entire planet was observed through $1\frac{1}{4}$ martian years. The long-term telescopic record, while meager, is also important, for the

one Viking year observed by MAWD may not be entirely representative of the present epoch (Jakosky and Barker, 1984).

The thin, cold martian atmosphere can hold little water. Even though close to saturation, it contains only 1–2 × 10^{15} g, or the equivalent of 1–2 km^3 of H$_2$O, as compared with 13,000 km^3 in the Earth's atmosphere. If all the water precipitated out, it would form a layer only 0.7–1.4 × 10^{-6} m deep. Thus the present martian atmosphere is extremely dry. Because so little water is in the atmosphere, latent heating effects are small.

The variation of water vapor in the atmosphere as a function of the time of year during the Viking mission is summarized in Figure 2-1. At $L_s = 0$, the start of northern spring, the latitudinal distribution of water vapor was flat, ranging from 1 pr μm at high southern latitudes, where the seasonal CO$_2$ cap was beginning to form, to 10 pr μm at 25°N. Longitudinal variations are negligible; the numbers in Figure 2-1 are zonally (latitudinally) averaged means. When the northern high latitudes warmed, as we moved into northern spring and summer, water vapor in the northern hemisphere started to increase. The increase began in the 60–80°N latitude band, before exposure of the residual water cap, which suggests that the source of water was the regolith and/or the seasonal cap. At about $L_s = 120$, the northern seasonal cap completely evaporated. At this time temperatures at the pole rose from 150 K, the frost point of CO$_2$, to around 200 K, the frost point of water, and the water content of the atmosphere over the residual cap rose to 100 pr μm. Clearly, the CO$_2$ cap had evaporated to expose a residual water-ice cap (Kieffer et al., 1976; Farmer et al., 1976). Meanwhile, the water vapor abundances were rising in the 30–60° N latitude band and continued to do so after sublimation from the residual cap ceased at about $L_s = 130$. The midlatitude increase was probably due in part to southward transport from the residual cap,

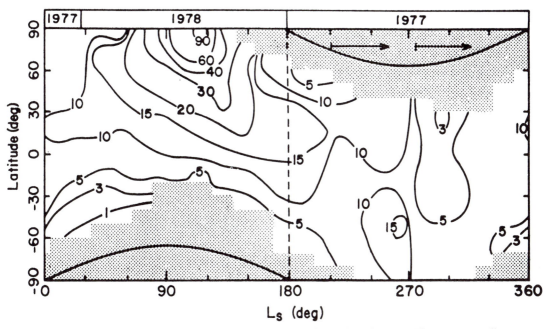

Figure 2-1. Column abundances of water vapor as a function of latitude and season. Figures are zonally averaged abundances in pr μm. Horizontal arrows show the dust storm periods. Poleward of the arcuate lines at high latitudes the Sun does not rise above the horizon. The most striking feature of the diagram is the increase in water vapor in the northern hemisphere during summer, particularly right at the pole when the water–ice cap emerges from beneath the receding CO_2 seasonal cap. (From Jakosky and Haberle, 1992. Reproduced with permission from the University of Arizona Press.)

but modeling of the circulation at this time suggests that latitudinal mixing is too sluggish to account for all the increase in water vapor observed (Haberle and Jakosky, 1990). Part of the increase was probably caused by movement of water out of the regolith. After $L_s = 140$, the water contents of mid to high northern latitudes fell as water became incorporated in the developing seasonal cap and resorbed into the regolith as it cooled. By the start of northern fall, at $L_s = 180$, the pulse of water from the residual cap and heating of the northern regolith had largely dissipated, and the latitudinal distribution was again rather flat. The second half of the year was less eventful. The southern seasonal cap did not completely disappear. No residual water-ice cap was exposed, and water abundances remain low. The 15 pr μm contour in Figure 2-1 at $L_s = 270$, the southern summer solstice, may indicate some loss of water from the regolith at mid to high southern latitudes, but it could also be an observational or dynamical artifact.

The behavior just described can be thought of in terms of interchange between several reservoirs, as shown in Figure 2-2. In northern spring and summer, water is transferred from the northern regolith and northern residual cap into the atmosphere of the northern hemisphere. Haberle and Jakosky estimate that in this process 0.1–0.8 mm of ice are removed from the northern cap. Some small fraction of the water injected into the atmosphere of the northern hemisphere is exchanged with the southern hemisphere, as suggested by the 10 pr μm contour in Figure 2-1. Water also condenses from the atmosphere onto the southern CO_2 seasonal and residual caps. During northern winter, water vapor in the atmosphere in the northern hemisphere recondenses on the seasonal cap and the residual cap, and is resorbed onto the regolith. Meanwhile, water continues to condense on the southern residual CO_2 cap, although the amounts are very uncertain.

Observations of springtime increases in water vapor well equatorward of the seasonal cap and the apparent inability of the northern residual cap to supply all the water observed in the atmosphere at high latitudes suggest that the regolith is acting as a reservoir for water. Within the re-

NORTHERN SPRING/SUMMER

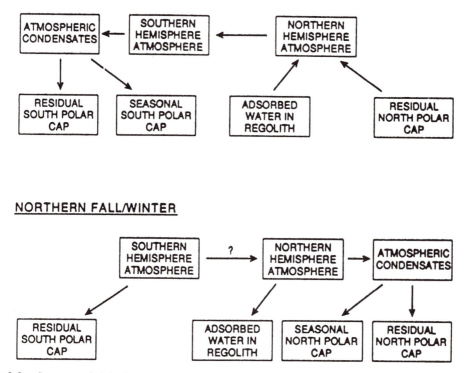

Figure 2-2. Sources and sinks for annually exchangeable water. (Jakosky and Haberle, 1992. Reproduced with permission from the University of Arizona Press.)

golith, exchange will take place between vapor, ice, and adsorbed water (Zent et al., 1986). At low wintertime temperatures, water will be driven into ice. In springtime the ice will tend to dissipate and the adsorbed component will tend to increase until temperatures rise sufficiently that both the ice and adsorbed components decline in favor of the vapor phase. The observational evidence for regolith-atmosphere exchange implies that the near-surface regolith high latitudes is porous, at least to depths of a few centimeters.

If the Viking year is typical of the present epoch and a residual CO_2 cap remains in summer at the south pole, then this acts as a semipermanent trap for water. This water cannot recycle through the atmosphere until the cap dissipates. However, the amount entering the residual cap must be small because it encompasses only a small area and the water vapor contents and gradients at high southern latitudes are very small. We do not know how typical the year of Viking observations was. Telescopic observations show-

ing similar water contents in northern and southern summers (Barker et al., 1970) suggest that in some years the southern seasonal cap may entirely disappear near the southern summer solstice and expose a water-ice cap. During 1969, for example, water contents during summer in the southern hemisphere rose to six times their Viking values, which suggests that the CO_2 cap disappeared and exposed water-ice (Jakosky and Barker, 1984).

A major issue with respect to the circulation just described is whether in the current epoch there is net, long-term transport of water from one hemisphere to the other. Unfortunately, the issue remains unresolved. The general north-south gradient in the water content of the atmosphere would seem to imply that net losses from the north are taking place. However, two effects may offset downgradient movement. Davies (1981) suggested that latitudinal transport in general, and the rate of atmospheric exchange between the two hemispheres in particular, depends on the dust content of the atmosphere. Exchange is minimal when the atmosphere is clear but vigorous when it is dusty. Thus, during northern summer, when the water content of the northern hemisphere is at its highest and the atmosphere is clear, circulation is sluggish and latitudinal transport is inhibited despite the strong water vapor pressure gradient. In contrast, in northern winter during the dust storm season, circulation is more violent and exchange is more vigorous. Water that has diffused away from the residual cap in summer is frozen back on the cap, and any loss of water from the northern to southern hemisphere is replaced. In the Davies model, there is no net exchange of water between the two hemispheres.

James (1985) suggested that the northern hemisphere might be the net gainer of water during the present epoch, despite the observed north to south gradient. Because of the orbital eccentricity, the south has long winters in which a more extensive seasonal cap develops than in the north. In contrast, the southern summers are short and hot. As a consequence, the off-cap winds due to sublimation of CO_2 are twice as strong in southern summer as in northern summer. The effects of this high south to north, southern summer flow more than offset the downgradient flow of water in the opposing season so that in James' model the southern hemisphere is becoming progressively drier and the northern residual water-ice cap is growing. The flow, of course, reverses with the precessional cycle.

In a third model (Jakosky and Haberle, 1990), in contrast to the two just described, there is a net annual flow from north to south. The modeling of summertime, high-latitude circulation indicates that it is too sluggish to transport much of the sublimed water away from northern residual cap so that most of it recondenses the following winter. The inability to transport the water vapor away from the cap results also in suppression of sublimation from the residual cap because of the high relative humidity. But according to this model, despite the sluggish circulation, there is a small net loss during the year from the residual cap. This, coupled with the possibility of a semi-permanent sink at the south pole, lead Jakosky and Haberle to conclude the northern hemisphere is slowly losing water to the south. We thus have three models: one in which the two hemispheres

are in balance, one in which the north is gaining water, and one in which the south is gaining water. Whichever way the net flow is, it is likely to be small.

Stability of Ground Ice

The thermodynamic stability of ice in the ground is controlled by the vapor pressure of water at the surface and by ground temperatures. If the ground temperatures are above the frost point of water, ground ice will sublime and water will be lost to the atmosphere by diffusion of water vapor through and out of the soil. If temperatures are below the frost point, water in the pores of the soil will freeze and water vapor will diffuse from the atmosphere into the soil. The stability is quite sensitive to the frost point temperature. The partial pressure of water at the surface can be readily derived from the water vapor density at the surface, given by

$$M(T) = \rho_0\,(T) \int_{o}^{h} e^{\,(-z/H)}\,dz, \qquad (1)$$

where M is the total normalized mass of water in the vertical column, ρ_0 is the density of water vapor at the surface, h is the height to which water is mixed, z is height, and H is the scale height of the water. From the partial pressure of water at the surface, the frost point can be derived as shown in Figure 2-3. For a completely mixed atmosphere containing 10 pr μm of water, the frost point temperature at the surface is 196 K. For an atmosphere containing 15 pr μm of water with mixing restricted to the lower 5 km, the frost point temperature is 203 K. These are reasonable limits on the expected frost point temperature.

To assess the stability of water-ice under present-day martian conditions, Farmer and Doms (1979) assumed a frost point of 198 K (a well-mixed atmosphere containing 12 pr μm of water) and average martian values for the albedo and thermal inertia, then determined where temperatures below the surface exceeded the frost point. In these places, water-ice will tend to sublime and be lost to the atmosphere. Where temperatures never reach the frost point, ice is permanently stable under present conditions and water will tend to diffuse from the atmosphere into the soil. The results are shown in Figure 2-4. Three regions are defined. At low latitudes, mean tem-

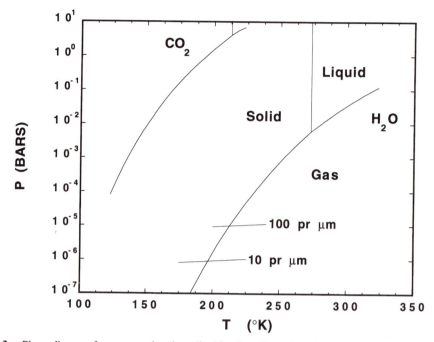

Figure 2-3. Phase diagram for water and carbon dioxide. Crosslines show the water vapor pressure at the surface for a well mixed atmosphere containing the amounts of water indicated. Frost-point temperatures are given by the point where these lines cross the solid–gas phase boundary.

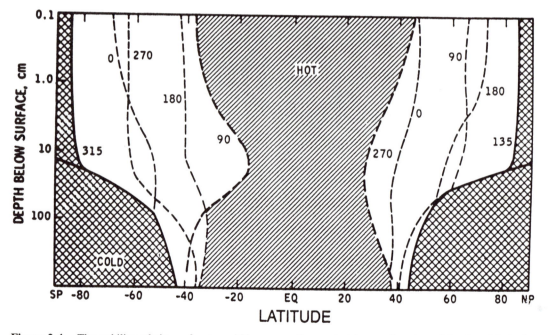

Figure 2-4. The stability relations of water within near-surface materials as a function of depth and latitude. The shaded region at high latitude shows where ice is always stable. At the poles, this region has been extended up to the surface because of observations of water-ice at the north pole. The shaded region at low latitudes is where ice is always unstable for the present obliquity. In the unshaded region ice is unstable part of the year. (From Farmer and Doms, 1979. Copyrighted by the American Geophysical Union.)

peratures are above the frost point, and temperatures at all depths below the effects of the diurnal wave remain above the frost point all year so that ice is permanently unstable at all depths. At the south pole, under the residual cap, ice is stable all year to depths of several kilometers. At the north pole, ice becomes unstable at the surface near midsummer, when it is permanently illuminated, and high sublimation rates occur. At all high latitudes ice is stable throughout the year at depths below about a meter (skin depth for the annual wave) because the mean annual temperature is below the frost point. At high latitudes, at a depth shallower than about 1 m, is a region where ice is stable parts of the year and unstable the rest of the year.

These stability conditions are of fundamental geologic importance. If present-day conditions are representative of much of Mars' geologic history, then in the martian low latitudes ground-ice is permanently unstable. If water was present in the soil when the present conditions were imposed, the soil at these latitudes must have become progressively desiccated with time, as any ice originally present would have tended to sublime and be lost to the atmosphere. Some of the losses may have been replenished by juvenile water or by water that has somehow re-entered the groundwater system. These possibilities will be explored later. Initiation of present-day climatic conditions probably initiated a desiccation wave that propagated into the planet. The loss rates of water and the propagation rate of the desiccation wave would have depended on the rate at which water vapor was able to diffuse through the overlying materials into the atmosphere. The water lost from the low-latitude terrains would have been frozen out at the poles and possibly become incorporated into the layered deposits there.

This somewhat simple picture of the stability relations of water-ice has been recently questioned by Paige (1992) and Mellon and Jakosky (1993, 1995). Farmer and Doms (1979) assumed a well-mixed atmosphere and average values for albedo and thermal inertia. Paige points out that Mars has a wide spread in the values of albedo and thermal inertia, and that the atmosphere may not be well mixed, factors that could affect the stability of ice. Mean annual ground temperatures are lower for lower thermal inertia soils than for higher thermal inertia soils for otherwise identical conditions, because the higher maximum daily temperatures and the T^4 dependence of emitted radiation result in more loss of energy by emission. Thus, ground-ice should be stable at lower latitudes in the low thermal inertia regions of Tharsis, Elysium, and Arabia (Figure 1-6) than in high inertia regions elsewhere. Presence of ground ice will increase the thermal inertia below the surface, so that a plausible model for the near surface materials at mid to high latitudes is an ice-saturated, high thermal inertia layer overlain by an ice-free low thermal inertia soil at the surface. Low thermal inertia materials at the surface would also result in ice closer to the surface than in the "average" model. The presence of a high thermal inertia layer at depth would suppress temperature fluctuations there. Finally if this model is combined with a frost point temperature of 203°K instead of 198°K, then ice becomes stable almost everywhere (Figure 2-5).

Mellon and Jakosky (1993, 1995) elaborated upon Paige's model by taking the observed thermal inertias (Figure 1-6) and albedos, and calculating mean annual temperatures as a function of location. They then went on to calculate how the stability would change with obliquity (Figure 2-6). Mean annual temperatures are lower in Tharsis and Arabia as expected and ice is more stable than elsewhere at the same latitudes. Although the low latitude surface temperatures change only a few degrees with small changes in obliquity, the stability of ice changes dramatically, mainly because of changes in the frost point temperature. At obliquities above about 32°, the models of Mellon and Jakosky indicate that the frost point will rise such that ice will be stable everywhere at shallow depths. Thus, while the Farmer and Doms stability relationships are valid in general for present climatic conditions and average surface properties, there are many exceptions. The stability of ice depends on the local albedo and thermal inertia, and slight changes in obliquity can have significant effects.

Megaregolith Holding Capacity for Groundwater and Ground Ice

Abundant geologic evidence for water erosion indicates that there are, or were, substantial amounts of water close to the martian surface, despite the presence of only minute amounts in the two reservoirs that we have observed directly,

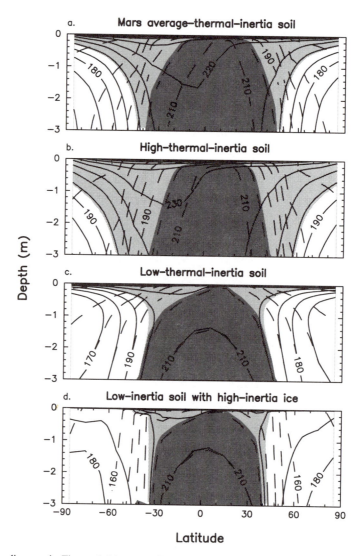

Figure 2-5. The diagram in Figure 2-4 is generalized. The stability of ice depends on local soil properties as shown here. Annual minimum (dashed lines) and annual maximum (solid lines) temperatures are shown as a function of latitude and depth for different soils. In the dark areas ice is unstable throughout the year, in the white areas it is stable throughout the year. Ground ice should, therefore, be found at shallower depths and lower latitudes in low thermal inertia soils than in high thermal inertia soils, particularly if sufficient ice is present to affect the thermal inertia below the surface. (Adapted from Paige, 1992. Courtesy of D. Paige, UCLA.)

the atmosphere and the north pole residual cap. If Mars has retained significant amounts of water, it must be hidden in the ground, either as groundwater or ground ice, and be inaccessible to direct observations. In this section we discuss the stability relations of liquid water and ice beneath the martian surface, and the potential holding capacity of the near-surface materials for water and ice. The discussion is based mainly on a series of papers by Clifford and his coworkers on the sta-

bility relations of groundwater and ground ice on Mars (Clifford, 1987, 1993; Clifford and Hillel, 1983; Clifford and Fanale, 1985).

The capacity of the martian crust for holding water can be estimated by analogy with the Moon. The numerous large-impact craters in the martian uplands suggest that the uplands were highly brecciated to substantial depths as a result of repeated impacts during heavy bombardment. We also see evidence in the uplands for volcan-

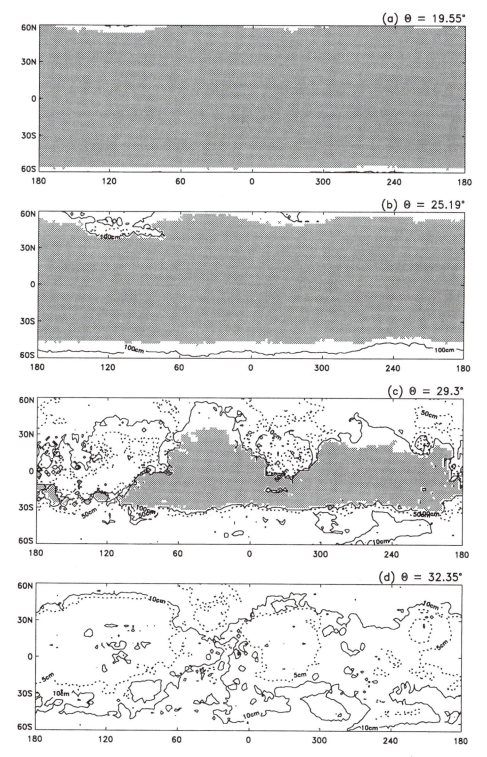

Figure 2-6. Models of the stability of ground ice as a function of obliquity. The different panels show stability as a function of latitude and longitude for different obliquities, Θ. In the shaded areas ice is unstable at all depths. Contour lines in the unshaded areas show depths below which ice is stable. At the present obliquity (panel b), ice is unstable at all depths at latitudes less than about 50°. At lower obliquities the stability boundary moves slightly poleward. At higher obliquities the stability relations change dramatically such that at obliquities higher than 30° ground ice is stable at shallow depths everywhere on the planet. (Adapted from Mellon and Jakosky, 1995. Courtesy of M. Mellon, NASA/Ames.)

ism, erosion, and sedimentation. Thus, the near-surface materials in terrains that have survived since heavy bombardment likely consist mostly of impact ejecta, intercalated with volcanics and sediments, which grade downward into a highly fractured basement of indeterminate origin, but which probably formed in the late stages of accretion (Figure 2-7). Younger terrains must be underlain at some depth by similar materials. A comparable model has been suggested for the Moon. There, the fractured, blocky, impact-stirred layer that formed during heavy bombardment has been termed a *megaregolith*. On the Moon, the highly fractured, porous materials that constitute the megaregolith have been confirmed to substantial depths from the seismic properties of the crust (Toksoz et al., 1974). The porosity of the megaregolith is expected to decrease with depth as a result of self-compaction. The seismic data indicate that the transition from porous megaregolith to coherent, essentially continuous basement as a result of closure of the pore spaces is nearly complete at a depth of about 20 km on the Moon, where the pressure is roughly 1 kbar.

To explain the seismic properties of the lunar crust, Binder and Lange (1980) suggested a model in which the porosity $\Phi(z)$ at depth z is given by

$$\Phi(z) = \Phi(0)\exp(-z/K_p), \qquad (2)$$

where $\Phi(0)$ is the porosity at the surface and K_p is the porosity decay constant with depth. Clifford (1993) applied a similar model to Mars. He assumed that the porosity decay constant would vary inversely with gravity and derived a martian decay constant of 2.82 km from the lunar decay constant of 6.5 km and the gravities on Moon (1.62 m s^{-2}) and Mars (3.71 m s^{-2}). Using this decay constant and integrating from the surface down to where the porosity had fallen to only 1%, he calculated that the total holding capacity of the martian megaregolith is 540 m and 1.4 km of water spread over the whole planet for surface porosities of 20% and 50%, respectively.

It is difficult to know what the appropriate porosity to use for Mars is. The decay law just given applies to near-surface rocks, not the loose surficial debris. While the soils analyzed by the

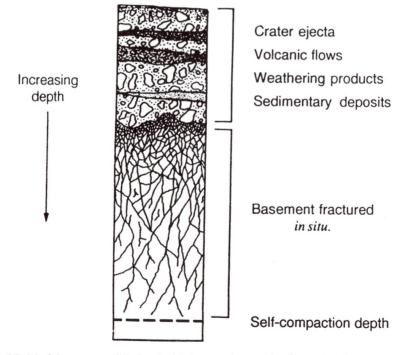

Increasing
depth

Crater ejecta
Volcanic flows
Weathering products
Sedimentary deposits

Basement fractured
in situ.

Self-compaction depth

Figure 2-7. Model of the megaregolith. Interbedded crater ejecta, volcanics, and sediments rest on a basement that is highly fractured as a consequence of large impacts during heavy bombardment. The self compaction depth, where the porosity has fallen below 1%, is estimated to be at a depth of roughly 10 km. (Reproduced by permission from Clifford, 1981.)

Viking landers had porosities of about 50%, they are likely to be representative only of a thin, wind-deposited veneer on a rocky substrate with a significantly lower porosity. Application of the decay law best fits lunar seismic data for a 20% porosity at the surface. Because the processes of megaregolith generation are similar on Mars and the Moon, the assumption of similar surface porosities for the two bodies is reasonable. On the other hand, the above-mentioned estimates are too low if substantial amounts of ice exist as discrete bodies rather than being restricted to the pore spaces. Discrete bodies of ice are to be expected on several grounds. On Earth, freezing of water-rich ground leads to segregation of ice into discrete lenses (see Chapter 5), and terrestrial permafrost commonly contains in excess of 50% ice. Moreover, during heavy bombardment if water was present at the surface, as appears likely, lake ice and other forms of ice could have been incorporated into the megaregolith by burial. In addition, the porosities of the deeper martian megaregolith could have been enhanced with respect to those on the Moon, by the presence of water under high hydrostatic pressures. Thus arguments can be made for both the 20% and 50% models.

Under present martian conditions, with mean annual temperatures well below freezing, the near-surface materials are frozen to form a thick permafrost zone. Only below the permafrost zone would we normally expect to find liquid water. In the literature on Mars the term *permafrost* is frequently used erroneously. The permafrost on the Earth is defined strictly on the basis of temperature. It is that part of the ground where temperatures have not risen above 273 K for 2 years. Ice need not be present in the zone. In the martian literature, *ground ice* and *permafrost* are commonly used interchangeably, thereby leading to confusion.

Clifford (1993) estimated the part of the megaregolith that is permanently frozen under present climatic conditions, and following Rossbacher and Judson (1981) called this permanently frozen near-surface zone the *cryosphere.*[*] The depth to the base of the cryosphere (z_c) is given by

$$z_c = K \, (T_{mp} - T_{ms})/Q, \qquad (3)$$

where K is the thermal conductivity of the cryosphere, Q is the geothermal heat flux, and T_{mp} and T_{ms} are, respectively, the melting temperature of ice and the mean annual temperature at the surface. Only T_{ms} is well defined. The other parameters must be estimated.

Clifford (1993), in reviewing the measurements of the thermal conductivity of frozen soils and basalts, likely components of the megaregolith, suggested that the best estimate for thermal conductivity is 2.0 W m^{-1} K^{-1}. Noting that the groundwater would likely be saline, he suggested a 252 K temperature at the base of the cryosphere, based on the work of Clark and Van Hart (1981) and Brass (1980). From various published models of the thermal evolution of Mars, he adopted 30 mW m^{-2} for the present heat flux (see Figure 1-7), as compared with 82 mW m^{-2} for the Earth and 16 mW m^{-2} for the Moon. Given these values and the mean annual temperatures as a function of latitude, the depths to the base of the cryosphere as a function of latitude are readily calculated from equation 3 (Table 2-1). The nominal depths range from 2.3 km at the equator to 6.5 km at the poles. Clifford also calculated maximum and minimum depths by combining extreme values of all the parameters in equation 3. For example, the cryosphere would have zero thickness at low latitudes in the extremely unlikely event that the freezing point was depressed to 210 K by exactly the right combination of salts. The maxima and minima given by Clifford are very unlikely, and although formal errors cannot be calculated, deviations of more than ±50% from his nominal values are probably rare. The cryosphere thickness must have increased with time as the heat flow declined. Schubert and Spohn (1990), for example, estimated that the heat flow at the end of heavy bombardment was around 150 mW m^{-2} (see Chapter 1). This effect alone would have resulted in a thickness of only 0.5 km for the permafrost zone at the equator at the end of heavy bombardment, assuming present climatic conditions.

Having estimated the depth to the base of the cryosphere and the porosities within it, Clifford (1993) was then able to estimate the storage capacity of the cryosphere. For the nominal model with a 20% surface porosity, its storage capacity

Table 2-1. Latitudinal Variation of Cryosphere Thickness.

(Adapted from Clifford, 1993)

Latitude	Mean Annual Temperature, K	Depth, km		
		Minimum	Nominal	Maximum
±90	154	1.24	6.53	23.8
±80	157	1.18	6.33	23.2
±70	167	0.96	5.67	21.2
±60	179	0.69	4.87	18.8
±50	193	0.38	3.93	16.0
±40	206	0.09	3.07	13.4
±30	211	--	2.73	12.4
±20	215	--	2.47	11.6
±10	216	--	2.37	11.3
0	218	--	2.27	11.0
Thermal model parameters.				
Geothermal heat flux (mW m^{-2})		45	30	15
Melting point temperature (^{0}K)		210	252	273
Thermal conductivity (W m^{-1} K^{-1})		1.0	2.0	3.0

is the equivalent of 374 m of water spread evenly over the whole planet. The total storage capacity of the megaregolith in this model is 544 m, leaving a storage capacity of 177 m below the cryosphere, where if water were present it would be liquid. For the 50% porosity model, the storage capacity of the cryosphere is 940 m, leaving 460 m of storage below the cryosphere. Finally, knowing the surface relief Clifford was able to estimate the height of the water table below the cryosphere for different quantities of water in the aquifer system. Figure 2-8 shows different levels of water within the global aquifer system for the 50% porosity model, assuming that the water table conforms to a geopotential surface, as it would, given time, interconnectivity, and no recharge from above. The figures just given are all for the present day. Obviously, earlier in the planet's history, when heat flow was higher, the cryosphere would have occupied a smaller fraction of the megaregolith.

Loss of Ground Ice to the Atmosphere

The discussion just concluded concerned the holding capacity of the megaregolith. It placed rough limits on the amounts of water that could be stored within the megaregolith but said nothing about how much might actually be present. The amounts in the cryosphere and the underlying global aquifer system depend on the total near-surface inventory of water and on the stabil-

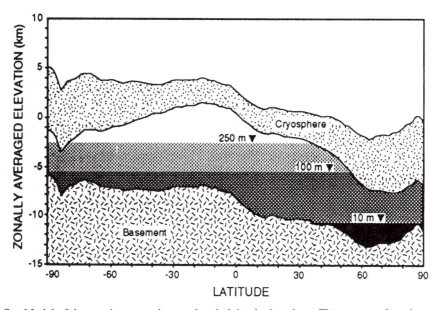

Figure 2-8. Model of the martian cryosphere and underlying hydrosphere. The upper surface shows the latitudinally averaged mean elevations. The surface of the basement is 10 km below the ground surface. The thickness of the cryosphere is taken from the values shown in Table 1-1. The water capacities shown are for the 50% surface porosity model. (From Clifford, 1993. Copyrighted by the American Geophysical Union.)

ity relations outlined earlier. We have seen that at low latitudes under present climatic conditions, ground ice, and at greater depths groundwater, is unstable with respect to the present atmosphere because ground temperatures are above the frost point. Any ground ice present will tend to sublime and the water will be lost to the atmosphere. The rate of loss will depend on the rate at which water vapor can diffuse through the overlying materials into the atmosphere.

Calculations of the loss rates of ice from the megaregolith are very uncertain. First, we do not know what the initial distribution of ice was. Because of the abundant evidence for water erosion, particularly early in the planet's history, it is reasonable to assume that at the end of heavy bombardment, ground ice and groundwater were widely distributed. Second, we do not know what surface conditions were like at the end of heavy bombardment, nor how they have changed since. Variations in the amount of CO_2 in atmosphere would have affected surface temperatures (Pollack, 1979), and variations in the amount of H_2O in the atmosphere would have affected the frost-point temperatures. Third, and perhaps most crucially, we do not know the properties of the megaregolith. As already indicated, the megare-

golith is likely to be very inhomogeneous, with impact ejecta interbedded with volcanics and sediments. There are likely to be weathering horizons and zones of cementation, formed as a result of leaching and precipitation from groundwater (Soderblom and Wenner, 1978; Burns, 1988). The duricrust, observed at the Viking landing sites (Mutch et al., 1977; Binder et al., 1977), for example, is likely to have a significantly lower permeability than the uncemented soil. Thin, low permeability layers, such as compact clays or cemented zones, would act as diffusion barriers for water vapor and invalidate calculations based on generalized properties. Nevertheless, it is useful to try to quantify the diffusive losses to get a sense of the efficiency of the process.

The following is a brief simplification of diffusive loss models by Clifford and Hillel (1983) and Fanale et al. (1986). Water vapor escapes from the soil by diffusion through the CO_2 in the pores. If the mean free path of H_2O in the CO_2 is less than the pore diameter, molecular diffusion dominates. If the mean free path is greater than the pore diameter, flow is governed by Knudsen diffusion. From experimental data the molecular diffusion coefficient of H_2O through CO_2 can be approximated by the expression

$$D_{AB} = 0.1654 \, (T/273.15)^{3/2}(1.013/P), \quad (4)$$

where T is the absolute temperature and P is the pressure in bars (Wallace and Sagan, 1979). The Knudsen coefficient for straight cylindrical pores is given by

$$D_{KA} = (2r/3)(8TR/\pi M_A)^{1/2}, \quad (5)$$

where r is the radius of the pores, R is the gas constant, and M_A is the molecular weight of water (Clifford and Hillel, 1983). These two coefficients can be combined into an effective diffusion coefficient D_{eff}, where

$$D_{eff} = D_{AB}D_{KA}/(D_{AB} + D_{KA}). \quad (6)$$

For the very low concentrations of water vapor in the host CO_2 gas, the flux of water vapor through the pores (J_A) of a soil can be approximated by

$$J_A = \frac{D_{eff}\,p}{kT\tau} \, \frac{dP_A}{dz}, \quad (7)$$

where p is the porosity of the medium, P_A is the vapor pressure of water, k is Boltzman's constant, and τ is the tortuosity, which, following Smoluchowski (1968), both Clifford and Hillel (1983) and Fanale et al. (1986) assume to have a value of 5. Equation 7 can be solved iteratively by standard numerical techniques to model the movement of water vapor through a soil profile as the P_A and T change in the profile.

Because of the dependence of D_{eff} on pore size, in each step in the numerical iteration the flux must be summed over all the pore sizes in the soil. Clifford and Hillel (1983) took a variety of pore size distributions that had been deter-

mined for terrestrial soils and calculated how long it would take for a 200 m thick layer of ice 100 m below the surface to evaporate. They found that for present-day mean surface temperatures at low latitudes (215–225 K), about half the soils would have lost all their ice in the last 3.8 billion years. In the other half the desiccation wave did not reach to the bottom of the ice layer. In similar but more elaborate calculations, Fanale et al. (1986) determined the depth of the desiccation wave as a function of time and latitude for soils with uniform pore radii. They included in their model the possible effects of obliquity variations on surface temperatures, surface pressures, and the water content of the atmosphere. Typical results are shown in Figure 2–9. For the more plausible 10 μm pore size model, the desiccation wave penetrates to a depth of a few hundred meters in 3.8 billion years, a result entirely consistent with the results of Clifford an Hillel.

The water lost by sublimation would ultimately have been trapped at the poles. At each pole is a thick sequence of layered deposits, which are suspected to be mixtures of water-ice and dust (Cutts, 1973; Thomas et al., 1992). Fanale et al. (1986) estimated that 2.8–5.6 × 10⁶ km³ of water have been removed from the equatorial near-surface materials since the end of heavy bombardment, a volume that is comparable to the estimated volume of 7 × 10⁶ km³ for the polar layered terrains. This estimate of the amount of water removed is probably an upper limit. The volume of pore space in the megaregolith down to a depth of 200 m in the ±30° latitude band, assuming a porosity of 0.2, is 2.8 × 10⁶ km³. Thus the entire megaregolith to this

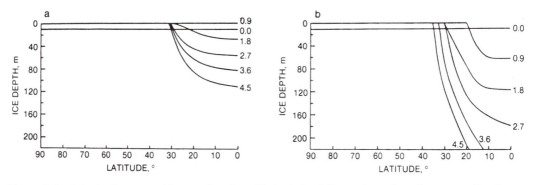

Figure 2-9. The depth to ground ice as a function of latitude for different times since the present climatic conditions were initiated and for assumed pore size of (a) 1μm and (b) 10 μm. (From Fanale et al., 1986. Reproduced with permission from Academic Press.).

depth must have been originally completely saturated with ice to achieve the estimated losses. Moreover, the layered terrains may have gained water by other processes, such as volcanism and catastrophic flooding.

These models are highly idealized. Nevertheless, the story seems clear. If ice was present in the megaregolith at the end of heavy bombardment, as seems likely from the abundant evidence of water erosion, most ice at low latitudes is likely to have been lost down to depths of a few hundred meters as a result of sublimation and diffusion into the atmosphere. This conclusion is probably valid even if there has been significant excursions in the climate that caused equatorial temperatures to be raised a few tens of degrees and allowed more water in the atmosphere, for the effects of increased frost-point temperatures resulting from higher water content of the atmosphere would be offset by the increased surface temperatures. Loss of ice may, however, have been suppressed during periods of high obliquity when ground ice may have been stable everywhere on the planet (Figure 2-6).

Although most of the equatorial ice has probably been lost, local survival of primitive ice deposits, where protected by diffusive barriers, may be common. Moreover, if in mid to late martian history climate changes have occurred that were so large as to allow precipitation at the surface, as has been suggested (Baker et al., 1991), then ice could have been replenished from above and be abundant at low latitudes. In addition, some replenishment may have occurred from below, as a consequence of outgassing of the interior and circulation of groundwater.

Migration of Water Through the Megaregolith

Clifford (1993) suggested three processes whereby water lost from the cryosphere could be replenished by water from below. The first process is thermal vapor diffusion. This results from the logarithmic dependence of the vapor pressure of water on temperature. Because of the geothermal gradient, the vapor pressure of water increases with depth at a faster rate than the increase from barometric relations. Thus, if there is a water table at some depth below the cryosphere, there will be a vapor pressure gradient

and water vapor will diffuse upwards. As it rises, temperatures will fall, and with them the vapor pressure. As a result, water will condense and the water droplets will tend to seep back down toward the water table. If impermeable zones are present, liquid water could be trapped well above the basal water table to form perched aquifers. If the base of the cryosphere is saturated, a rough equilibrium will be maintained between the rising vapor and the returning liquid. Clifford (1993) calculated that for a temperature gradient of 15 K km^{-1}, roughly the present thermal gradient, the flux of vapor at the base of the cryosphere is 2.8 $\times 10^{-4}$ m yr^{-1}, or roughly 1 km in 4×10^6 years. The rate would have been higher earlier in the planet's history when heat flows were three to five times higher. Thus the potential for thermal vapor diffusion to resupply a depleted cryosphere is large.

The second process that Clifford suggested could resupply the cryosphere is seismic pumping. Shock waves generated by impacts and earthquakes can produce transient dilatation and compression of water-bearing formations and force water toward the surface. Clifford estimates that impacts alone have resulted in over 10,000 events larger than the 1964 Alaskan earthquake, which resulted in eruption of water from shallow aquifers onto the surface at distances as far as 400 km from the epicenter. Formation of the large basins, such as Argyre and Hellas, could have disrupted the cryosphere and vented aquifers to the surface on a global scale.

The third process for vertical water transport is hydrothermal circulation. Hydrothermal activity can result wherever groundwater is available and a thermal anomaly is present. The possibility of volcanically induced hydrothermal activity has been discussed by several authors, including Squyres et al. (1987) and Gulick (1992). Hydrothermal activity could also result from thermal anomalies created by impacts (Newsom, 1980; Brackenridge et al., 1985). Circulation results as water, heated by the anomaly, rises and is replaced by lateral migration of water from around the anomaly. The rising water tends to cool and flow radially away from the anomaly so that local circulation may be established. If heating is sufficient, hot water and steam would create pathways through the cryosphere and cause hot springs and venting of steam at the surface.

Hydrothermal activity is thus of special biologic interest (see Chapter 9). Given that Mars has been volcanically active throughout its history, if the crust is water rich, which is likely, then hydrothermal activity has probably been common. However, although hydrothermal activity may have been efficient in locally circulating groundwater, it is doubtful that water vented into the atmosphere or introduced onto the surface in springs could very readily be returned to the ground under the present climatic conditions, for even the most vigorous hydrothermal system is likely to provide only very local areas where ground temperatures right at the surface are above freezing.

Of the three processes just described, thermal vapor diffusion has the greatest potential for resupplying the cryosphere from the water table below, provided water can be transported through the cryosphere and not be choked off at its base. Water can move through unsaturated frozen ground by water vapor diffusion much in the same way that it moves through unfrozen ground. However, the upward-moving vapor will be precipitated as ice in the pores of the soil as the temperatures fall during the ascent. In this case there is no return flow, as in unfrozen soil. The soil will ultimately become saturated, cutting off further vapor diffusion flow.

Water can still pass through ice-saturated soils. Significant amounts of liquid water may be adsorbed on grains in frozen soils down to temperatures well below freezing, particularly in the presence of freezing point depressors such as $NaCl$ and $CaCl_2$ (Anderson and Tice, 1973), but as the temperature declines so does the content of adsorbed water. The pressure or hydraulic head of the soil water, in the presence of ice, is inversely proportional to temperature. A temperature gradient causes a pressure gradient; therefore, the adsorbed water will move through the frozen soil in a manner loosely analogous to the way water moves through any permeable medium. However, because the movement is restricted to the surfaces of grains, the hydraulic conductivity drops very rapidly as temperatures drop below freezing and the fraction of unfrozen water drops (Williams and Smith, 1989).

Water also can move through a saturated frozen soil by a poorly understood process called *regelation*, which involves pressure-induced melting and freezing of ice in the pores. At temperatures a few degrees below freezing, all the particles within a saturated frozen soil are encased in an unfrozen film of adsorbed water. The rest of the pore space is filled with ice. Regelation is a counterintuitive process in which a fall in water pressure on the cold sides of pores causes melting and an increase in pressure on the warm side causes freezing. Water released on melting diffuses down the thermal gradient, along grain boundaries and across grain contacts, to be frozen onto ice filling an adjacent pore. The result is a net movement of ice from the warm to cold side of pores (Williams and Smith, 1989), and transport of water through the saturated soil.

The two latter processes are likely to be effective only at the base of the cryosphere, where temperatures are close to freezing, because the amount of adsorbed unfrozen water in frozen soils drops so dramatically with temperature (McGaw and Tice, 1976; Gooding et al., 1992). Anderson et al. (1967) found, for example, that at temperatures 10°C below freezing adsorbed water in a bentonite represented only a monolayer around the grains. For lower temperatures the amount present is even less. Thus migration of water along films on the grain boundaries, and transformation of ice to water and back again, would be an extremely slow process at temperatures more than a few degrees below freezing. Because vapor diffusion transport in frozen soils is self-limiting, it is difficult to see how large fractions of ice could build high in the martian cryosphere as a result of transport from below.

We have just been discussing vertical transport of water, but water will move laterally through a permeable medium if there is a hydraulic gradient. Clifford (1993) suggested that the global, interconnected megaregolith provides a medium whereby groundwater can be redistributed on a global scale. In particular, he suggests that the equatorial regions could be resupplied with water via the megaregolith to compensate for diffusive losses to the atmosphere. Much of the water that has entered the atmosphere as a result of processes such as diffusive loss from the megaregolith, sublimation from lakes and streams, volcanic activity, impacts, and so forth must have ultimately been frozen out at the poles. The layered sediments at the poles are thought to be mixtures of ice and dust, and the main repository for this water (Pollack et al., 1979; Murray et al., 1972).

The northern polar deposits are sufficiently thick (Dzurisin and Blasius, 1975) that melting could occur at their base, particularly early in the planet's history, when heat flows were higher. The meltwater would seep into the megaregolith, creating a pole-to-equator hydraulic gradient. A circulation could thus be established in which the water table is lowered at low latitudes as a result of losses to the cryosphere by thermal vapor diffusion. The cryosphere loses water to the atmosphere, and this is frozen out at the poles. Ice is buried at the poles until it reaches such a depth that it melts. The meltwater trickles down to the water table and flows out radially from the poles back to the equator. Clifford found that a global aquifer a few kilometers thick with a permeability comparable to the Earth's crust (10^{-2} Darcy's) could transport the equivalent of a global ocean a few hundred meters thick from the poles to the equator over the course of martian history. The rate-controlling element in the global cycle described by Clifford may not be groundwater flow but some other component of the circulation, such as diffusive losses from the deeper cryosphere at low latitudes.

It should be re-emphasized that most of the discussion in this chapter pertains to present climatic conditions. If conditions were very different at times in the past, as has been suggested, then the circulation of groundwater, replenishment of the equatorial aquifer system, and interchange between the atmosphere and the megaregolith would be greatly facilitated.

Summary

The water cycles on Earth and Mars are very different (Figure 2-10). High surface temperatures on Earth result in a largely moist atmosphere. Any redistribution of water by the atmosphere is efficiently counteracted by runoff and movement of ocean water. The only significant reservoir of water other than the oceans is polar ice, and this constitutes only 2.8% of the total. Trapping of water as ice at the Earth's poles is counteracted by calving of ice back into the oceans and equatorward flow of cold, bottom water. The martian atmosphere, in contrast, contains only minute amounts of water. The bulk of any water present on the planet must be within the surface as ice at shallow depths and as liquid water at greater depths. The holding capacity of the megaregolith is high, with estimates ranging from 0.5 to 1.4 km spread over the whole planet, but we do not

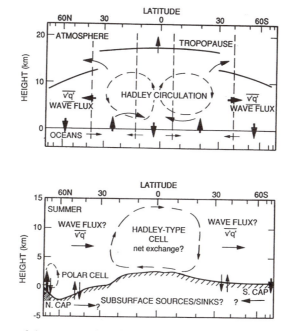

Figure 2-10. Comparison of the water cycles of Earth (upper) and Mars (lower) (From Zurek, 1992. Reproduced by permission from the University of Arizona Press.)

know what fraction is actually filled. Exchange of small amounts of water take place annually between the caps, the high latitude regolith, and the atmosphere, and the surface materials at low latitudes are probably very slowly losing water to the atmosphere. Any net additions to the exchangeable water inventory are counteracted by trapping at the poles. It is uncertain whether at the present time one cap is losing water to the other. Also uncertain is whether water-rich deposits at the poles are undergoing, or have undergone in the past, basal melting so that water is returned to the groundwater system, thereby allowing it to migrate to lower latitudes.

3 OUTFLOW CHANNELS

The term *outflow channel* was coined by Sharp and Malin (l975) to denote large channels, several to tens of kilometers across, that start full size at discrete sources. Tributaries are rare but branching downstream is common. Outflow channels have bedforms on their floors, sinuous streamlined walls, enclose teardrop-shaped islands, and tend to be deeper close to their sources than they are downstream. Bedforms include longitudinal scour, inner channel cataracts, and plucked zones. Many, particularly those that debouch into the Chryse-Acidalia basin, expand when they cross nearly level plains to scour ill-defined areas up to hundreds of kilometers across. The terms *channels* and *valleys* have genetic implications and misuse can lead to confusion. A *fluvial channel* is the conduit through which a river flows and is at times filled or almost filled with water. A *valley* is a general term that refers to any linear depression. A *fluvial valley* generally contains many fluvial channels, is normally orders of magnitude wider and deeper than the channels it contains, and never comes close to being filled with water. Despite their large sizes, the outflow channels of Mars appear to be true channels, cut by enormous floods.

Distribution and Characteristics

Outflow channels occur in four main areas: around the Chryse-Acidalia basin, centered at 20°N, 45°W; in Elysium Planitia, centered at 30°N, 230°W; in the eastern part of Hellas basin, around 40°S,270°W; and along the western and southern margins of Amazonis Planitia, which is centered at 20°N, 160°W (Figure 3-1). By far the largest concentration of channels is around the Chryse basin. The basin is a sparsely cratered plain almost completely surrounded, except to the north, by cratered uplands. Elevations on the plain are mostly at least l km below the Mars datum. Outflow channels start around Chryse, either in the cratered uplands or in high volcanic plains, at elevations as high as 4 km above the datum. They then extend downslope, and converge on Chryse Planitia from the east, south, and west (Figure 3-2). On the Chryse plains, they broaden to scour wide swaths of terrain and ultimately disappear northwards at about the 40°N latitude.

Many of the channels around Chryse basin start in areas, termed *chaotic terrain*, where the ground has seemingly collapsed to form broad areas of jostled blocks that may be as much as 1–2 km below the surrounding, undisturbed terrain (Figure 3-3). Many individual blocks retain evidence that they were formerly at the surface, indicating that the chaotic terrain was formed by collapse, not by removal of material from above. Channel walls commonly merge with the walls enclosing the chaotic terrain, and linear depressions may be partly filled with rubble, so that the distinction between channel and chaotic terrain is often blurred. Most of the chaotic terrain, and hence the source of the channels, is an area extending from 20°S to 5°N, and 15°W to 50°W, at the east end of the canyons. Others channels start in box canyons, containing chaotic terrain, to the north of the main canyon system. Many of the low-lying areas of chaotic terrain to the east of the canyons merge westward with the canyons, and northwards with outflow channels, so that the walls enclosing all three components are continuous. Not all chaotic terrain is directly connected to a canyon or a channel. Several large, completely enclosed, areas indicate that the void into which the surface materials collapsed was formed by massive removal and transport of material entirely below the surface.

Some of the circum-Chryse channels do not start in chaotic terrain. Thick sedimentary sequences in several parts of the canyons, particularly in the Ophir, Candor, Hebes, and Ganges Chasma, suggest the former presence of deep

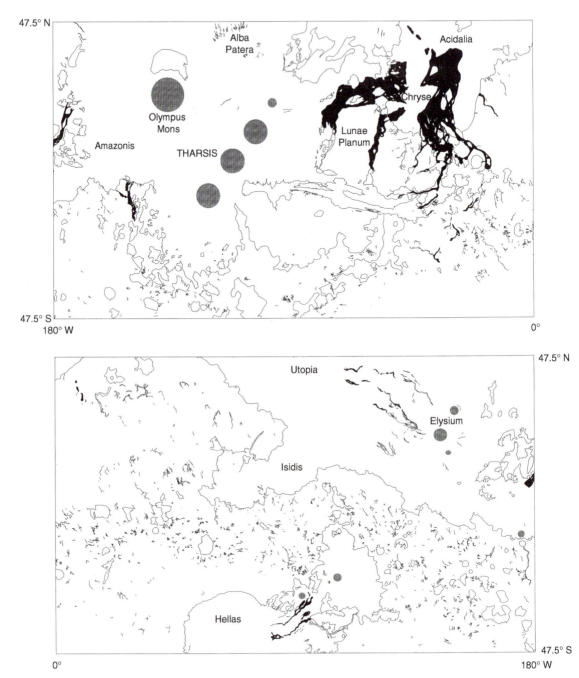

Figure 3-1. Distribution of outflow channels in the ±47.5 latitude band. (a) Western hemisphere. (b) Eastern hemisphere. The outflow channels, shown in black, drain mainly into four areas: Amazonis-Arcadia, Chryse-Acidalia, Hellas, and Utopia. Volcanoes are indicated by shaded circles, whose size and location are only approximate. The volcano Alba Patera is left unshaded so as not to obscure the portrayal of valleys on its flanks. The canyons, chaotic terrain, and the Hesperian-Noachian boundary are also shown to indicate location. Valley networks are also shown.

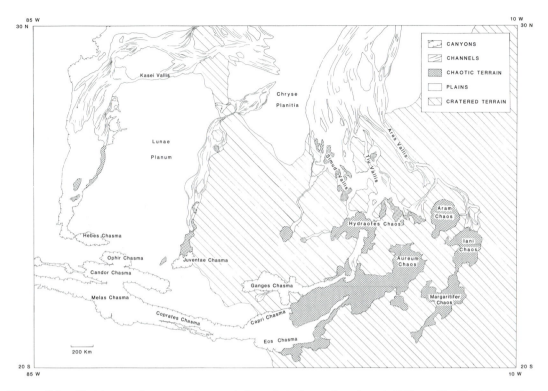

Figure 3-2. Sketch map of canyons, chaos and channels to the south and west of Chryse Planitia. Most of the channels entering Chryse Planitia from the south start in chaos regions or emerge from the east end of the canyons. Kasei Vallis, the largest outflow channel, enters Chryse Planitia from the west and starts in a shallow north-south depression, north of the main canyons. (From Carr, 1981.)

lakes (McCauley, 1978; Nedell et al., 1987; Lucchitta et al., 1992). Sculpted, streamlined forms in Capri Chasma and Eos Chasma at the eastern end of the Valles Marineris have been taken as evidence of catastrophic release of the ponded water, downslope to the east. The postulated lake in Ganges Chasma may have partly drained northward by subsurface flow to form Shabaltana Vallis, as indicated in Figure 3-4, which shows that a linear depression, into which surface materials have collapsed, extends almost from the rim of Ganges Chasma, northward to an area of chaotic terrain from which Shabaltana Vallis emerges. Possibly the largest of all the outflow channels, Kasei Vallis, starts in Echus Chasma, a broad, north-trending, flat-floored depression, situated north of the Valles Marineris, Echus Chasma. Robinson and Tanaka (1990) propose that the channel formed by catastrophic drainage of a lake in this depression.

The channels themselves, where they cut through the uplands, are mostly broad, flat-floored, steep-walled valleys. The walls commonly follow gently sinuous paths and may be terraced. One of the most distinctive attributes of the channels is the presence of streamline islands (Figure 3-5), with pointed prows upstream and long tapering tails downstream. Many of the islands form where flow has been diverted by some obstruction, such as a crater. Flow around the islands may be traced by deep scour marks in the channel floor (Figure 3-6), or by a depression or moat that wraps around the island. In many places, deep grooves, several hundred meters across and some several tens of meters deep, have been eroded into the floors of the channels (Figures 3-3 and 3-6). Some reaches have grooves that are gently sinuous and subparallel; in other reaches the striae diverge and converge in complex patterns. In the upper part of Kasei Vallis, the scoured, grooved swath reaches widths of 200 km. The scour marks, being bedforms, and the islands demonstrate unequivocally that the channels are true channels, formed by

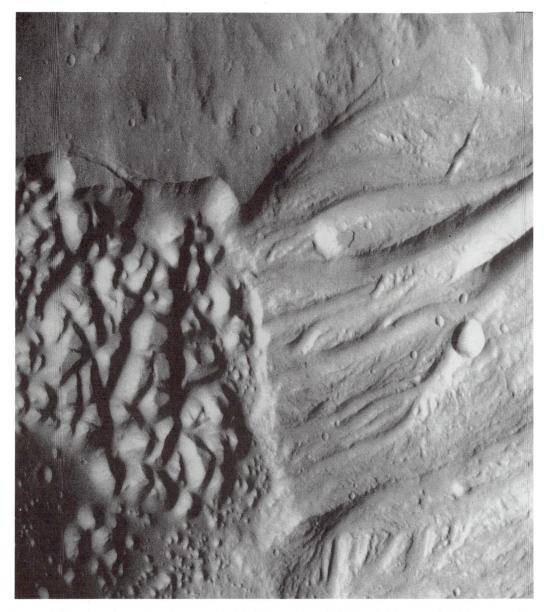

Figure 3-3. Channel emerging from a rubble-filled depression at 20°N, 33°W. The scene is 62 km across. The deep scour on the channel floor is indicative of the great depth of the flood (Viking Orbiter frame 742A12).

catastrophic floods, and not valleys, formed by slow erosion.

The channel shape, the streamlined islands, and the striated floors are the most distinctive features of the outflow channels, but many other features have close analogs in terrestrial flood. These include inner channels, cataracts, and etched or plucked zones. For a complete description of the characteristics of the martian outflow channels and comparisons with large terrestrial

floods, see Baker and Milton (1974), Baker (1982), and Baker et al. (1992). Depositional landforms such as deltas have not been clearly identified.

The large outflow channels in the Elysium region originate in a manner very different from those around the Chryse basin. Most originate on the west and northwest flanks of the Elysium rise, in graben-like depressions radial to the rise. To the northwest, these rectilinear depressions

Figure-3-4. Ganges Chasma and its environs. Ganges Chasma, in the lower half of the picture, is 150 km across at it widest. A thick sequence of layered sediments form an "island" within the northern half of the canyon. To the north of the canyon a vague depression, into which surface has partly collapsed, leads from the canyon to an area of chaotic terrain at the top of the picture. Out of this terrain emerges a large channel, Shalbatana Vallis, which is not in the picture. The relations suggest that water within a lake in Ganges Chasma drained northward, underground to the source of Shalbatana Vallis.

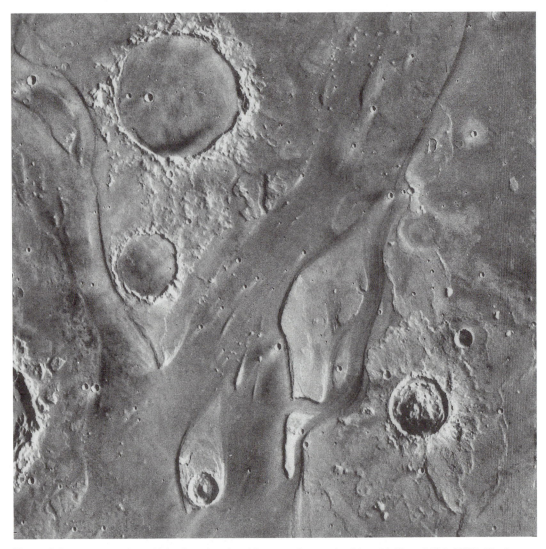

Figure 3-5.   Teardrop-shaped islands and sculpted forms at the mouth of Ares Vallis at 20°N, 33°W. The largest crater is 62 km across (Viking Orbiter frame 827A21).

become sinuous and fluvial-like, and teardrop-shaped islands, like those described in the circum-Chryse channels, become common. The fluvial-like forms are generally found only at elevations below the 0 km contour. The channels can be traced from the graben, several hundred kilometers to the northwest, where they change from well-defined channels, with flat floors and steep walls, to vague, sinuous dark markings with little discernible relief.

The rectilinear Hephaestus Fossae in southern Elysium Planitia are different from all other outflow channels, if indeed they are outflow channels. A sinuous channel emerges from an irregu-

lar depression. Downstream the channel branches and the individual branches become first rectilinear and then discontinuous (Figure 3-7). If formed by water, much of the erosion (or solution) must have occurred below the surface, preferentially along lines of structural weakness. The pattern is very similar to that in terrestrial karst (Palmer, 1990, for example), where solution of limestone is the dominant conduit-forming process. If formed by subsurface solution and erosion, they are unlikely to have formed catastrophically like most of the other channels discussed in this chapter.

The channels in eastern Hellas (Figure 3-8) re-

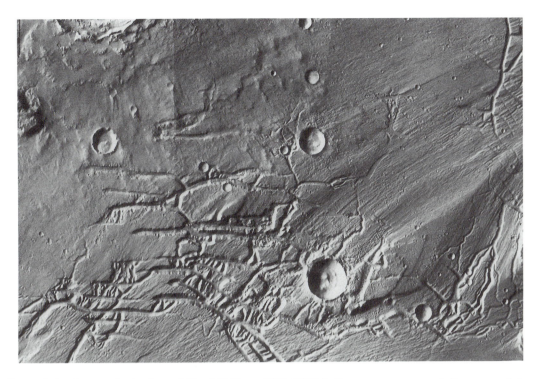

Figure 3-6. Grooves in the floor of Kasei Vallis at 27°N, 71°W indicate convergence and divergence of the flow. Coarse transverse crevasses suggest plucking along linesof weakness. The largest crater, near the center of the image is 7 km across. (From Robinson and Tanaka, 1990.)

semble those around Chryse. They start full size and extend down the regional slope into the floor of the Hellas basin. The largest, Dao Vallis, starts close to the large, old volcano Hadriaca Patera. Only two outflow channels of any appreciable size debouch into Amazonis Planitia. Mangala Vallis emerges from a graben (Figure 3-9) and extends northward through the cratered uplands for several hundred kilometers until it crosses the plains-upland boundary, beyond which it cannot be traced. It is the best photographed of all the outflow channels and displays a rich array of erosional forms, including terraced walls, teardrop-shaped islands, scoured floors, branching and rejoining channels, and expansions and contractions. Finally, a broad, shallow channel can be traced from 5°N, 190°W in southern Elysium Planitia to 23°N, 175°W in Amazonis Planitia. As can be seen in Figure 3-10, this channel has almost no superimposed impact craters. It is the youngest outflow channel so far recognized. Scott et al. (1992) suggested that it formed by drainage of a former lake in southern Elysium Planitia.

The outflow channels have a wide spread in ages, but the vast majority of them are Hesperian in age (Masursky et al., 1977; Scott and Tanaka, 1986). They mostly formed after the end of heavy bombardment, but still early in Mars' history. The main exceptions are the channel in Figure 3-10 and some of the channels northwest of Elysium, which are thought to be Amazonian in age (Greeley and Guest, 1987).

Origin of Outflow Channels

The channels of Mars were first recognized during the *Mariner 9* mission (Masursky, 1973). Because it was understood at that time that liquid was unstable at the martian surface under present climatic conditions, there was some reluctance initially to accept these features as fluvial in origin, and alternative modes of formation were examined. The alternatives included erosion by lava (Carr, 1974; Schonfeld, 1977), liquid hydrocarbons (Yung and Pinto, 1978), wind (Blasius and Cutts, 1979), liquid CO_2 (Sagan et al., 1973), glaciers (Lucchitta, 1982) and liquefaction of the

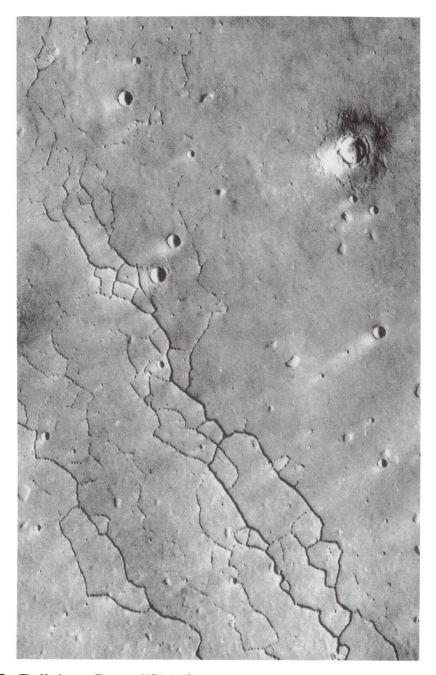

Figure 3-7. The Hephaestus Fossae at 22°N, 239°W. This network of depressions merges to the southeast into a single depression that starts abruptly at a local source, much like other outflow channels. However, this feature is clearly not formed by flow across the surface. The pattern is similar to those found in terrestrial karst and formed by solution of limestone by subsurface streams (Viking Orbiter frame 381S13).

surface materials (Nummedal, 1978). None of these alternatives, however, appears as plausible as water erosion. Lavas can erode, as evidenced by lunar and venusian sinuous rilles, but there is no supporting evidence for volcanism at the sources of the largest of the channels, those around the Chryse basin. Moreover, lavas erode by thermal incision and generally form simple channels. Once the channel is well incised, it is difficult for the lava to overflow, so lunar rilles

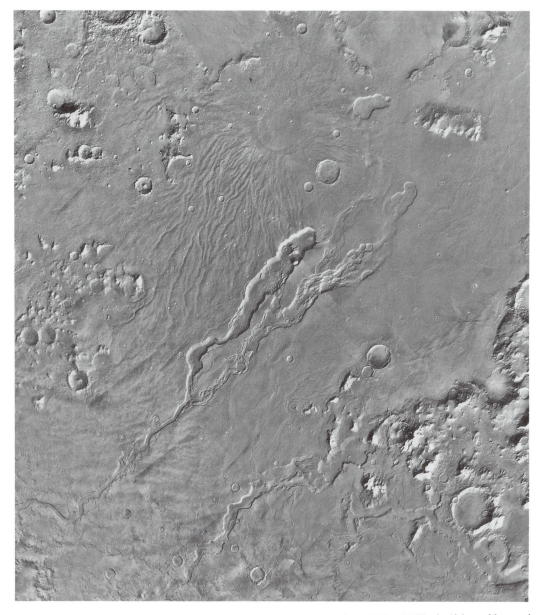

Figure 3-8. Large channels in eastern Hellas. The most prominent channel, Dao Vallis, is 40 km wide near its source. The channel transects ridges crudely radial to the volcano, Hadriaca Patera, whose center is faintly visible at the top center of the picture. The channels extend southwestward into the floor of Hellas, which is to the southwest of the bottom left corner of the picture. Volcanic activity may have triggered formation of the channels by melting of ground ice and by thinning the permafrost, thereby providing easy access of groundwater to the surface.

rarely branch, and are very unlike the channels just discussed (Carr, 1974). Scoured floors, in particular, are unlikely to be produced with lava.

The liquid hydrocarbons and liquid CO_2 hypotheses have problems with the source, release, fate, and properties of the fluids involved. Wind is also a very unlikely erosive agent. Surface winds tend to be coupled to winds aloft. The at-

mosphere does not flow downhill, following the intricacies of local topography, and focusing its erosive power in single, well-defined channels, like those observed. The outflow channels resemble glacial valleys in many ways, and although an ice sheet fed by groundwater, could conceivably spawn glaciers, there is no supporting evidence for such sheets in the source regions of the

Figure 3-9.   The source of Mangala Vallis at 16°S, 149°W. The channel originates at a graben and continues northward for several hundred kilometers. Groundwater may have been able to reach the surface because of disruption of the permafrost seal by the graben. Alternatively, or additionally, the faults may have provided pathways for groundwater flow. The picture is 250 km across (Viking Orbiter frame 639A11).

circum-Chryse channels, nor are there moraines downstream. Moreover, there is clear evidence of subsurface erosion (Figures 3-4 and 3-7, for example), which is impossible with a glacier.

Finally, liquefaction may have played a major role in the collapse of the surface to form the chaotic terrain, but dense, liquified slurries are unlikely to have travelled and maintained their erosive power for the hundreds of kilometers required to cut the large channels. Because of these

and similar arguments, and because of the resemblance of the outflow channels to large terrestrial flood features, liquid water is now almost universally regarded as the erosive agent that formed the outflow channels (for summaries see Carr, 1981; Baker, 1982).

The outflow channels closely resemble large terrestrial flood channels, such as those of the Channelled Scablands of eastern Washington (Baker, 1973, 1982; Baker and Milton, 1974). The

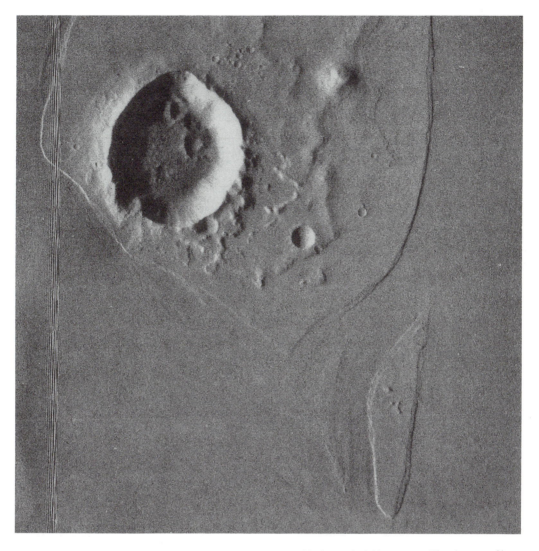

Figure 3-10. Outflow channel in western Amazonis Planitia. The image is 34 km across. The absence of impact craters superimposed on the channel floor indicates a young age (Viking Orbiter frame 690A44).

Channeled Scablands are among the largest known terrestrial flood features. They formed at the end of the Pleistocene by episodic catastrophic release of water behind an ice-dammed lake, Lake Missoula, that covered parts of Idaho and western Montana in the western United States. The floods surged down the Columbia River valley, overflowed the valley, stripped thick deposits of loess from the basaltic bedrock, and carved deep channels both in the basalt and the overlying loess. Flood depths are well defined by high-water marks on loess islands. The resemblance between the Lake Missoula floods and the martian outflow channels, is quite striking. Both typically have sections where erosion is re-stricted to deep, sinuous channels, with scoured floors and sections where broad poorly channelized swaths of terrain were scoured. Both enclose teardrop-shaped islands. In both, channels may form branching and rejoining patterns, and the pattern of scour indicates expansion and constrictions in the flow. Indeed, the resemblance between the two is so strong, and there is so much other supportive evidence for water erosion, that a flood origin for most of the outflow channels can hardly be doubted.

The two most probable causes for the martian floods, massive release of groundwater and draining of surface lakes, have already been mentioned. As will be shown later, discharges of the martian floods may have reached peaks as high as 10^9 m^3 s^{-1}. For comparison, the average discharge for the Mississippi is 2–3 × 10^4 m^3 s^{-1}, and peak discharges of the Channeled Scabland flood and the Chuja Basin flood in Siberia are estimated to be 1–2 × 10^7 m^3 s^{-1} (Baker, 1973, Baker et al. 1993). Thus, some of the martian floods may have been almost 100 times as large as the largest known terrestrial floods.

The large floods could have formed as a result of massive release of groundwater under high artesian pressures confined below the thick permafrost zone (Carr, 1979). A thick permafrost, and hence climatic conditions similar to those at present were probably necessary for the mechanism to work. With a thin permafrost zone, or no permafrost zone, the required pressures may not be able to build because of leakage of water with the surface. The high artesian pressures could be caused simply by the generally low elevations of the source regions (most are less then +1 km) compared to the water level in the regional aquifer system. Alternatively, or additionally, if the aquifer system were saturated hydrostatic pressures would build early in the planet's history as the permafrost zone thickened in response to the falling heat flow or changing surface conditions. Breakout from the aquifer could have been triggered by any event that disrupted the permafrost seal. The most likely trigger is a large impact, either by penetrating the permafrost seal or by sending a large pressure pulse through the aquifer, such as happens in terrestrial aquifers following earthquakes. Another possibility is breakage of the permafrost seal by movement along a fault, as is suggested by the origin of Mangala Vallis at a graben (Figure 3-9).

As the flood progressed, the rate of flow would have declined as the local artesian pressure dropped and the flood would have been over when water could no longer reach the surface. The aquifer would then have resealed by freezing and the process could have repeated if the aquifer were recharged by flow from parts of the aquifer distant from the flood. I originally calculated the aquifer permeabilities required to provide the discharges estimated for the outflow channels (Carr, 1979). I now believe that these calculations were in error. The discharges from the aquifer would have been so large that the aquifer would have been disrupted and the host rock materials carried away in the flood. The flow would therefore not be constrained by the permeability. Collapse of the surface to form the chaotic terrains is testament to the removal of the aquifer materials.

Proximity of channel sources to volcanoes, as in Elysium and Hellas, may result because of the higher thermal gradients in these regions. The higher gradients would cause melting of ice in the cryosphere, thereby providing a local source of liquid water and, by thinning the cryophere, providing easy access of water to the surface. Groundwater might preferentially move along fault planes, so emergence of channels from graben, as we see in Elysium and with the Mangala Vallis, is not surprising. Massive movement of groundwater along faults radial to Tharsis may be responsible for subsurface erosion and/or solution in the canyon regions, as is suggested by lines of pits and closed depressions along the radials (see Figure 1-11), and particularly by the emergence of channels from some of these pits at the east end of the canyon.

The possible origin of some outflow channels by catastrophic release of water in lakes within the canyons, as suggested by McCauley (1978), has already been mentioned. Given a global aquifer system, it is to be expected that water would pool in the canyons. The Ophir, Candor, and Melas Chasma in the central section of the Valles Marineris are over 8 km deep (Lucchitta et al., 1994). This is the deepest part of the Valles Marineris, but most of the canyons are over 4 km deep. At present, the nominal thickness of the cryosphere is 2.3 km at the equator (see Table 2-1), but it was probably thinner at the time the layered sediments (Figure 3-11), which are the main evidence of former lakes, accumulated. Thus, the canyon floors were well below the expected base of the cryosphere (Figure 3-12), and water could have leaked into the canyons both while they were forming and after they formed. Under present climatic conditions, such ponded water would tend to stabilize at the level of the local water table to form an ice-covered lake, as discussed later in this chapter. Subsurface seepage out of the lakes, as suggested by Figure 3-4, for the Ganges Chasma might ultimately have led to undermining and catastrophic release of the ponded water and formation of some outflow channels.

Figure 3-11. The completely enclosed Hebes Chasma, which is 120 km across at its widest. An "island" within the canyon is composed of layered sediments, suggesting the former presence of a lake. The lake may have drained by sub-surface flow into the adjacent Echus Chasma in the upper left (Viking Orbiter frame 682A27). Compare with Figure 3-4.

Discharge Rates and Sediment Transport

Discharges for the martian floods have been estimated from empirical relations for terrestrial rivers (Komar, 1979). Ignoring the effects of the atmosphere, the retarding stress at the base of a flow, τ_r, is given by

$$\tau_r = \rho ghS, \tag{1}$$

where ρ is the density of the fluid, g is the gravitational acceleration, h is the depth of the flow, and S is the slope. This is equal to bottom stress, τ, created by the flow given by

$$\tau = \rho C_f u^2, \tag{2}$$

where C_f is a dimensionless friction coefficient and u is the mean flow velocity. Equating (1) and (2) gives

$$u = (ghs/C_f)^{1/2}. \tag{3}$$

The following empirical Manning equation is often used to predict flow velocities in terrestrial channels and rivers (Sellin, 1969):

$$u = (1/n) \, h^{2/3} \, S^{1/2}. \tag{4}$$

The Manning roughness coefficient, n, has been determined empirically for different kinds of natural and artificial channels, and tables are listed in Komar (1979). But because of the empirical nature of the data and the qualitative description of the types of rivers for which the coefficients were measured, application of equation (4) involves some geologic judgement. By combining (3) and (4) the roughness coefficient can be related to the Manning coefficient:

$$C_f = g(n^2/h^{1/3}). \tag{5}$$

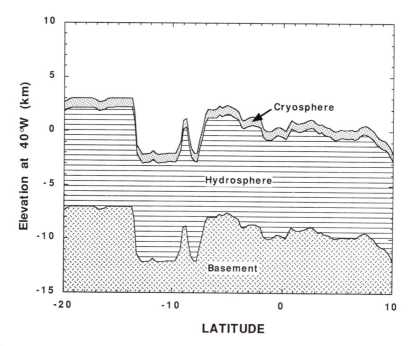

Figure 3-12. North-south section through the canyons at 44°W. The permafrost zone, estimated assuming the present climate and heat flows for early in Mars history, is labeled cryosphere. The diagram shows that the canyons were far deeper than the permafrost so that even with today's climatic conditions, liquid water below the permafrost could have leaked into the canyons creating lakes. The depth of the lakes would have been controlled by the local water table. The basement, drawn at 10 km below the surface as in Figure 2-8, is somewhat artificial and unlikely to perfectly follow changes in surface elevation as shown.

The appropriate C_f can then be substituted in equation (3) to determine the mean flow velocity, from which discharges can be calculated. In estimating the discharges for the Lake Missoula floods, for example, Baker (1973) used a Manning coefficient of 0.04 (dimensions expressed in meters), the value for a clean regular channel, with moderate sinuosity, thereby deriving a drag coefficient of 0.0034 for the 100 m thick flow he thought eroded the features observed.

One difficulty in determining the discharges for the martian floods is knowing the depth. Widths can be measured from the images, slopes can be roughly estimated from the available topography, and depths of the channels can be estimated from shadow measurements and stereo photography, where available. But the present depth of the channel gives only an upper limit for the depth of the flow that eroded it. Baker (1982) and Komar (1979) estimated that the floods that cut Maja, Ares, and Mangala Valles, if 100 m deep, had peak discharges, respectively, of 3×10^8, 7×10^7, and 2×10^7 m³ s⁻¹. A depth of 100

m is a reasonable guess. A lower limit is given by grooves on the floor (Figures 3-3 and 3-6), because the water must have been considerably deeper than the grooves, and these may be tens of meters deep. An upper limit is the depth of the channel where the width is measured, which varies considerably but is generally over 100 m. Discharge is proportional to $h^{5/3}$, so the figures just given can be scaled to other depths. Robinson and Tanaka (1990), suggested an upper limit for the Kasei Vallis discharges of $1.2–3.5 \times 10^9$ m³ s⁻¹. They were able to trace the same set of scour marks from the rim of the channel across the channel floor, and so were confident that their depth measurements were true stream depths. Kasei Vallis is considerably larger than the Maja, Ares, or Mangala, so a flood with a peak discharge approaching these high values is not unlikely.

Floods of such size must have had enormous erosive power and sediment-carrying capacity. From the size of Lake Missoula and the peak discharges implied by the high water marks in the

Channeled Scablands, Baker (1982) estimated that peak discharges during the Lake Missoula floods could not have been maintained more than a few days. Sediments left by the floods indicate that the floods were episodic, although few in number. During these short-lived events, the floods cut channels several kilometers wide and over a hundred meters deep in basalt. One of the most impressive creations of the floods, the Dry Falls cataract, 5.5 km across and 120 m high, probably formed by a few floods lasting only a few days. Baker (1982) described several phenomena that aided this erosion. Large vertical vortices or *kolks* produce enormous hydraulic lift along the filament of vortices, thereby plucking material from the bed. Horizontal vortices wrap around obstacles and trail downstream in their wake, focusing the erosive power of the flow. Flow velocities are so high in floods, particularly in the vortices, that the pressure commonly drops below the vapor pressure of water and bubbles develop that are carried downstream. Cavitation, or collapse of the bubbles, produces shock waves that add to the erosive capacity. A bedload of boulders, meters across, provides further erosive power. Most of these phenomena are only poorly understood.

Building on the work of Bagnold (1966) on sediment transport, Komar (1980) examined how sediments might be transported in these large floods. Material can be transported in a river in two basic ways, as a bed load, in which the sediment saltates or rolls along the bed, and as a suspended load, in which the grains are supported by the turbulence of the flow. The distinction between the two loads is blurred because of variations in the amount of turbulence and because grains with large saltating paths are in effect suspended. The coarser part of the suspended load is concentrated toward the bottom of the flow. The finest part of the suspended load is sometimes referred to as the *wash load*. The settling times of the particles in the wash load are long compared with the time scale of the flow, so it shows no vertical concentration gradient.

To distinguish between the bed load and the suspended load, Bagnold (1966) defined a dimensionless parameter k, which is defined as

$$k = w_s/u_*, \qquad (6)$$

where w_s is the settling velocity of the cut-off grain size between the bed load and suspended load and u_* is the frictional shear velocity given by

$$u_* = (\tau/\rho)^{1/2} = u(C_f)^{1/2}. \qquad (7)$$

Because of ambiguities in distinguishing between bed load and suspended load, estimates of the appropriate value of k vary. Komar (1980), after reviewing proposed values, used a value of 1.25. Combining equations (1), (6), and (7) we have,

$$w_s = k(ghs)^{1/2}, \qquad (8)$$

which gives the cut-off settling velocity for the suspended load. By relating settling velocities (w_s) to particle diameters by standard drag coefficient curves (Graf, 1971), Komar then was able to determine threshold diameters for the suspended load.

Komar (1980) strongly emphasized the importance of the wash load in martian floods. It has been found that a river expends no energy suspending particles with $w_s \leq us$. In fact, the material becomes autosuspended and adds to the power of the flow. Because the river expends no energy carrying the wash load, it can carry large fractions, and wash loads as high as 95% by weight have been measured. The wash load is normally not limited by hydraulics but by the availability of the fine-grained component in the materials being eroded. Little information is available on the cut-off size for the wash load in terrestrial streams. Komar (1980), after reviewing the available data, adopted k values of 0.05–0.1 for the threshold size. Again, these k values can be converted to particle diameters from drag coefficient curves.

Some of Komar's results are shown in Table 3-1. The threshold for the wash load is around 1–2 mm for the size floods likely on Mars, and Komar suggested that a large fraction of the sediment load could be in the wash load. If this were so, then the floods could carry large fractions of sediments. He thought that the martian floods might resemble flash floods in the southwest United States in which sediment loads of 60% by weight, or 40% by volume, are not uncommon. But whether such high wash loads would occur in the martian floods would depend on the fraction of 1–2 mm size and smaller grains in the material that was eroded. I am somewhat skeptical that the martian uplands consists dominantly of particles 1–2 mm in diameter or smaller, so am doubtful that the sediment loads were as high as Komar suggested.

Table 3-1. Threshold Values for Different Modes of Sediment Transport.
(After Komar, 1980)

	Bottom slope, S	Depth, h (m)	Shear, u* (cm/s)	Velocity, u (m/s)	D_t (cm)	D_s (cm)	D_w (cm)
Mangala Vallis	0.003	10	33	4.7	34	0.7	0.03
		50	75	10.6	170	4.2	0.07
		100	105	14.9	350	20	0.1
Ares Vallis	0.01	10	61	8.6	120	2.4	0.06
		50	136	19	460	80	0.13
		100	193	27	940	100	0.15

D_t = the threshold diameter for the bedload, D_s = the threshold diameter for the suspended load.

D_w = the threshold diameter for the wash load. Basalt densities are assumed (2.8 gm cm^{-3}).

The volumes of materials removed by the large outflow channels can be roughly estimated from the volumes of the landforms they produced. Table 3-2 from Carr et al. (1987) shows the volumes of various features around the Chryse basin. They were determined from a combination of shadow and stereo measurements. The biggest uncertainty in estimating the amount of material removed by water is knowing what fraction of the canyon volume was caused by faulting and what fraction by erosion. There is good evidence that part of the relief in the canyons was created by faulting (Blasius et al., 1977), but what fraction is unknown. By ignoring those parts of the canyon where faulting appeared to have dominated, Carr et al. (1987) suggested that around the Chryse basin roughly 4×10^6 km^3 of material had been eroded by water.

We can get some idea of how long the floods lasted by comparing the estimated discharges with the volumes of material removed. Floods are by nature short-lived events. If the source is a lake, the lake rapidly drains; if the source is groundwater, the artesian pressure rapidly drops. Ares Vallis drains the Margaritifer, Iani, and Aram Chaos (Figure 3-2). Combining their volumes with that of Ares Vallis from Table 3-2, we find that the flood(s) that caused Ares Vallis removed about 2×10^5 km^3 of material. If, follow-

ing Komar (1980), we assume 40% by volume of sediment, then the volume of water was 3×10^5 km^3. For a 100 m deep flood Baker (1982) estimated the peak discharge rate for Ares Vallis was 7×10^7 m^3 s^{-1}. Given this peak discharge, the total volume of water implies that the discharge decayed with a decay constant (time for discharge to fall to 1/e of the peak) of 50 days. If formed by a single flood 200 m deep, the decay constant would be 9 days. If the floods were multiple, the decay constants would be shorter. The latter number seems small for the size of the events we are discussing, but a decay constant of a month may not be unreasonable. The times vary inversely with the sediment fraction. If Ares Vallis was formed by multiple events then estimates would have to be changed correspondingly.

In summary, discharge rates from the large martian floods are estimated to be in the range of 10^7 to 10^9 m^3 s^{-1}, compared with 10^7 m^3 s^{-1} for the largest known terrestrial floods. Total amounts of water for individual floods would have ranged up to about 3×10^5 km^3 of water if they had very high sediment loads. The total amount of material eroded from around the Chryse basin is estimated to be 4×10^6 km^3, requiring at least 6×10^6 km^3 of water, and possibly considerably more. All these numbers are extremely uncertain. We do not know the depths of

Table 3-2. Volumes of features in the Chryse Region (in units of 10^4 km^3)

(From Carr et al. 1987)

Chasmata		Valles	
Echus	53.5	Kasei	36.9
Hebes	7.8	Maja	3.3
Ophir	12.7	Shabaltana	4.5
Candor	57.2	Simud-Tiu-Ares mouth	9.0
Tithonium	18.2	Ares	8.0
Ius	24.0		
Melas	91.2	Chaos	
Coprates	36.5		
Juventae	3.8	Pyrrhoe	2.1
Eos Capri	76.0	Hydroates	11.7
Ganges-Eos Junction	40.3	Hydaspis	4.5
Ganges	26.9	Iani	3.2
		Aram	2.8
Valles		Margaritifer	2.3
		Aureum-Arsinoes	13.2
Simud	21.8		
Tiu	18.2		

Total volume = 5.89 x 10^6 km^3

the floods, which are important for estimating discharges. The estimates of discharges are based on empirical relations for terrestrial rivers, and we do not know whether extrapolations of two to four orders of magnitude from the sizes of terrestrial rivers to the sizes of the martian floods are valid. Finally, the estimates of water volumes are based on extreme values for the sediment-carrying capacity of large floods, which also may be invalid.

Lakes in the Canyons

The main reason for thinking that the canyons formerly contained lakes is the presence in many parts of the canyons of thick sequences of rhythmically layered deposits. The thickest and most extensive deposits are in the Ophir, Candor, Melas, and Hebes Chasmata. At the boundary between Ophir and Candor, the sediments form a flat-topped, mesa-like construct with finely fluted walls that contrast markedly with the coarse, spur and gully topography of the partly buried divide between the two canyons. In the completely enclosed Hebes Chasma (Figure 3-11), the deposits form an 8-km thick sequence (Lucchitta et al., 1994) in the middle of the canyon and separated from the canyon walls by a deposit-free zone. Gaps also occur between canyon walls and the similarly thick sedimentary stacks in Ophir and Candor, so that the sediments are rarely seen in contact with the canyon walls, except at the divide between the two canyons. The horizontal layering is evident from variations in albedo of the different layers and from terracing of the walls in which the sequences are exposed. The total volume of deposits is substantial, being approximately 600,000 km³ (Lucchitta et al., l994).

Nedell et al. (l987) examined four possible modes of origin for the layered deposits: They were deposited by the wind, they are the same material as the canyon walls, they are ash deposits from explosive eruptions, or they were deposited in standing bodies of water. After examining the alternatives, they concluded, in agreement with McCauley (l978), that deposition from liquid water in the tranquil, low-energy environment of a lake provides the best explanation of the horizontality, lateral continuity, great thickness, and stratigraphic relations of the deposits. Additional support for the lacustrine hypothesis is the presence of fluvial features at the east end of the Valles Marineris, possibly caused by abrupt release of some of the ponded water,

and the abundant evidence of the former presence of groundwater in the region, as discussed earlier. However, there are many unsolved problems with the lacustrine hypothesis, including the source of the materials that formed the sediments, how they were removed from the source, and how they were transported to and deposited in the canyons. [See Lucchitta et al. (1994) for a general discussion of the formation of the canyons and the origin of the enclosed deposits and their surrounding troughs].

The >8 km thickness of the layered deposits suggests that if they were deposited in lakes, the lakes were long lived. We have just seen that formation of the large outflow channels by eruption of groundwater may have required a thick permafrost zone. Could lakes have been stable in the canyons with present climatic conditions, under which water would freeze and ice is unstable? The answer appears to be yes. Given extensive groundwater, as indicated by the outflow channels, and a permafrost zone that is considerably thinner than the depth of the canyons, water would tend to seep into the canyons and fill them to the level of the local water table (Figure 3-12).

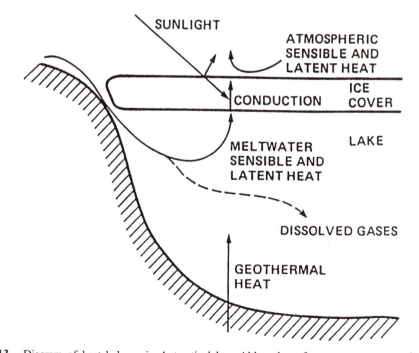

Figure 3-13. Diagram of heat balance in Antarctic lakes. Although surface temperatures rarely rise above freezing, the lakes do not completely freeze because of energy supplied by sunlight, meltwater, and geothermal heat. (From McKay et al., 1985. Reproduced with permission from Nature. Copyright 1985, Macmillan Magazines, Ltd.)

The water would freeze and develop a thick ice cover. If no water entered the lake, then the entire lake would freeze and, given time, the ice would ablate away. However, if groundwater were present, it would seep into the lake to compensate for ablation losses in order to maintain the lake level at the level of the local groundwater table. The ice would reach an equilibrium thickness, controlled in part by ablation losses at the surface. The situation is similar to that in ice-covered lakes in the Antarctic (McKay et al., 1985; Figure 3-13). Squyres (1989) estimated that under present conditions the ice cover of the proposed canyon lakes would be a few hundred meters thick at equilibrium. His estimates were based on clean ice. As indicated in the next section, ablation rates would be much less if the ice were covered with a few meters of lag or soil, and the equilibrium ice thickness would be correspondingly greater.

Terminal Lakes

In the low-lying areas at the ends of the outflow channels, the floods must have left large lakes in which their sedimentary loads were deposited. A considerable effort has been spent in trying to identify geomorphic evidence for such lakes (Jöns, 1985; Lucchitta et al., 1986; McGill, 1985, 1986; Parker et al., 1989, 1993; Scott and Chapman, 1991; Scott et al., 1992, 1995). Most of the effort has focused on the northern plains, in which the circum-Chryse and Elysium outflow channels terminate. The northern plains are complex, variegated, and poorly understood. They clearly have formed by a variety of processes. Over large areas the plains materials have been deposited over remnants of a highly cratered surface, which survives today mostly as arrays of low, roughly equidimensional hills or knobs that commonly outline large, ancient pre-plains craters. The plains materials themselves have probably formed by many processes, including volcanism, sedimentation from outflow channels and their resulting lakes, mass wasting from the adjacent uplands, and modification by impacts and by the action of ice. Ground ice is likely to be common at these high latitudes, and many of the features found there have been attributed to the action of ground ice (Carr and Schaber, 1977; Lucchitta, 1981; Rossbacher and Judson, 1981). The problems of identifying lacustrine features within this complexity are exacerbated by the generally poor photographic coverage of these regions and the subtlety of the lacustrine features being sought.

Because of these difficulties, questions of the size, fate, and effects of the former lakes are very controversial. Baker et al. (1991) suggested that the floods periodically caused global-sized oceans, possibly covering as much as a quarter of the planet's surface and temporarily changing the global climate. I, on the other hand, have argued for lakes comparable in volume to the estimates for individual floods, as discussed earlier in this chapter. These lakes would cover areas of no more than about 10^6 km^2, or 0.7% of the planet's surface, and would have had a trivial effect on the climate. Following is a brief discussion of some possible morphologic indicators of the former presence of lakes in the northern lowlands.

Polygonally fractured ground

Large areas of the northern plains are characterized by a polygonal pattern of fractures, with individual polygons ranging in size up to 20 km (Figure 3-14). The fractured plains occur mainly in two areas: in Acidalia Planitia, at the end of the Chryse channels, and in western Elysium and Utopia Planitiae, at the end of the Elysium channels. A genetic connection with the outflow channels, therefore, appears probable. Lucchitta et al. (1986) supported the suggestion of McGill (1985) that the polygonal patterns developed by contraction and compaction of sediments deposited from the channels.

Mottled plains

Many of the low-lying northern plains have a mottled appearance caused by contrast between bright crater ejecta and the intervening dark plains. Their location near the end of channels and distinctive erosional patterns suggestive of thermokarst (see Chapter 5) suggest that these mottled plains may be ice-rich deposits formed of frozen lakes.

Strandlines

Parker et al. (1989, 1993) have identified numerous linear features in the northern plains that they suggest are strandlines, or former shorelines. These included breaks in slope on escarpments

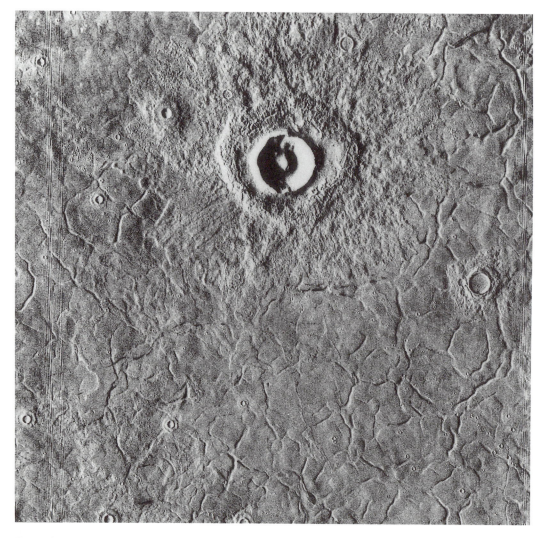

Figure 3-14. Polygonally fractured ground at 38°N, 258°W. Similarly fractured ground is generally found at the ends of outflow channels. The picture is 176 km across (Viking Orbiter frame 573A08).

and ridges within the plains (Figure 3-15), breaks in slope at the boundary between the plains and uplands, and boundaries between different textured units within the plains (Figure 3-16). They identify two discontinuous sets of strandlines. An inner set encloses an area of 27×10^6 km^2. An outer set runs roughly along the plains-upland boundary and encloses 46×10^6 km^2, roughly the area proposed by Baker et al. (1991) for their global oceans.

Terraced and stepped massifs

Many of the knobby remnants of the terrain underlying the plains are surrounded by low cliffs. Some, when viewed in high resolution, are seen to have terraced slopes. Parker et al. (1989), interpret both terraces and cliffs as wave-cut, and marking the former positions of shorelines, but other interpretations are possible.

Backflow features

Parker et al. (1993) point out several examples in the lower reaches of the circum-Chryse channels, where streamlining suggests flow in a direction opposite to the slope of the channel. Their interpretation is that filling of terminal lakes by floods would cause a rapid rise in the base level and backflow up the channels.

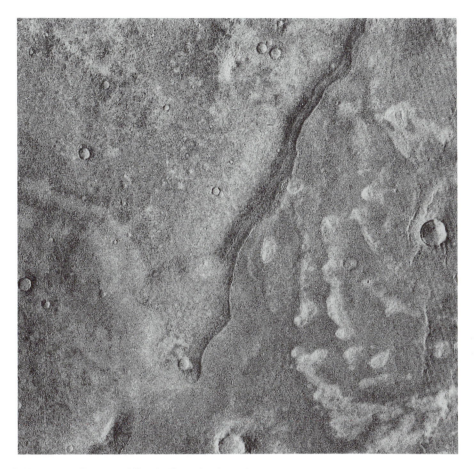

Figure 3-15. "Strandlines" at 46°N, 346°W. The linear features running diagonally across the scene have been interpreted by Parker et al. (1989) as shorelines of former lakes. The scene is 14 km across (Viking Orbitor frame 458B70).

Layered sediments

Costard (1988) has identified etched layered deposits in the low-laying northern plains that he plausibly interprets as lacustrine. The layers are well exposed in etched-out hollows that Costard suggests are thermokarst, that is, formed by sublimation of ground ice.

Because of the equivocal nature of much of the geomorphic evidence for lakes, it is difficult to assess how plausible some of the more extreme estimates for the sizes of the postulated lakes are. Scott et al. (1992), using criteria similar to those just listed, identified 15 different depositional basins on the planet with areas ranging as high as 3×10^6 km^2 for the Hellas basin. Some of the basins are connected by channels, and they suggest that some may have overflowed, and water

spilled over into an adjacent basin. On the other hand, Baker et al. (1991) postulate ocean-sized bodies of water, possibly covering areas as large as 3.8×10^7 km^2. They acknowledge that the ocean-sized lakes that they are postulating could not form from a single channel-forming event. They would require that many of the circum-Chryse channels formed simultaneously, one flood perhaps triggering another. Baker et al. (1991) show a sketch map of their proposed global ocean, on which the shoreline closely follows the 0 km contour. By their estimate, this would require 6.5 $\times 10^7$ km^3 of water, 10 times the estimate given earlier in this chapter, for the total volume of water required to cut the circum-Chryse channels. Not only is such a large body of water extraordinarily difficult to produce, it is difficult to eliminate. The proposal is discussed further in Chapter 6.

Figure 3-16. Mosaic of the boundary between two different types of plains at 46°N, 351°W. The boundary has been interpreted as a shoreline by Parker et al. (1989). The stippled unit with faint linear striations is on the lake side. Compare with Figure 5-12. The scene is 130 km across (Viking mosaic 211-5066).

Sublimation From Lakes

Under present climatic conditions, any terminal lake would have rapidly frozen. The subsequent fate of the ice would have depended on the latitude of the lake. We have seen that most of the channels end at high latitudes, either in the northern plains or in the floor of Hellas. At these latitudes, under present climatic conditions the mean annual temperature is below the frost point of water, so that if buried below the annual skin depth of 2–3 m, the ice would form a permanent, stable deposit. Lakes that pooled at low latitudes would tend to sublime away, because mean annual temperatures are above the frost point. Carr (1983, 1990) estimated sublimation rates and freezing rates for lakes under various conditions.

The heat balance at the surface of the lake is given by

$$-\epsilon\sigma T^4 + (1 - A)S + Q_c - Q_s + I = 0, \quad (9)$$

where ϵ is the infrared emissivity, usually assumed to be 1, σ is the Stefan-Boltzmann constant, T is the surface temperature, A is the albedo, S is the solar radiation incident on the surface, Q_c is the heat conducted to the surface from below, Q_s is the heat lost by sublimation, and I is the long-wave radiation incident on the surface, which, following Kieffer et al. (1977), is taken to be 0.02 times the incident short-wave radiation at noon. From Sellers (1965),

$$S = S_M (r/r_m)^2(\sin \gamma \sin \delta + \cos \gamma \cos \delta \cos h). \quad (10)$$

S_M is the solar constant at Mars, r_m and r are the mean distance to the sun and the distance at a particular time, γ is the latitude, δ is the solar declination, and h is the hour angle. Q_s has two terms, a turbulent water vapor flux that occurs as wind blows across the surface, and a convective flux, which occurs because water vapor is lighter than CO_2. For details on how these may be calculated, see Toon et al. (1980). Equation (9) can be solved iteratively by stepping through the year in discrete time steps, appropriately changing the incident radiation given by equation (10) and solving for temperatures in the ice at each iteration. The ice-water interface is kept at 273K, and heat losses from the underlying water are calculated at each iteration. These heat losses are compensated for by freezing an appropriate amount of ice onto the ice-water interface. Thus the growth of the ice is modeled, as are temperatures at the surface and the rates of sublimation.

The calculations suggest that at low latitudes under present climatic conditions the ice would thicken to over a meter in about a month and to 5–10 m in about a year, depending on the conditions. Sublimation rates are low because of the low temperatures on the surface of the ice and the logarithmic dependence of the vapor pressure of water on temperature. After the thermal anomaly created by the flood had dissipated, estimates are that at the equator sublimation rates would stabilize at around 0.01–0.1 cm yr^{-1}. Even though these sublimation rates are low, a 10-m thick ice deposit would sublimate away in a time that is geologically short, 10^4 to 10^5 years (Carr, 1990).

The figures just listed are for bare ice. More likely, dust would build on the ice and a lag would accumulate as the dirty ice sublimed. The cover so formed would impede sublimation in two ways. First, the soil would protect the ice surface from large daily and annual excursions in temperature. This has a major effect because of the strong dependence of water vapor pressure on temperature. Second, water sublimed from the top of the ice would have to diffuse through the soil into the atmosphere, and humidity in the soil would hinder sublimation. Sublimation rates under these conditions, estimated using equations (2)–(5) (Carr, 1990), show that at the equator with a 1 m thick cover, sublimation rates stabilize at about 10^{-5} cm yr^{-1}, so that 100 m of ice would sublime away in 10^9 years. These results are consistent with those of Fanale et al. (1986) for the stability of ground ice.

At latitudes greater than 30°, sublimation rates depend on the obliquity. At soil depths thicker than 1 m and obliquities less than 30°, the sublimation rates fall to negligible rates. Only at high obliquity does significant sublimation occur at high latitudes, and even this is at the very low rates of 10^{-5} to 10^{-6} cm yr^{-1}. A 100 m thick ice deposit at latitudes greater than 30° would take longer than the age of the planet to sublime. We should therefore expect to find ice deposits at the mouths of the large channels if climatic conditions similar to those at present have been continuously maintained.

If this reasoning is correct, then the presence of strandlines at high northern latitudes is very puzzling, for they imply that lakes were formerly present and have since evaporated. Baker et al. (1991) would argue that the sublimation calculations are irrelevant because they assume present climatic conditions. They suggested that eruption of the large quantities of water required for their global ocean would have dramatically changed the global climate, so that any calculation based on present conditions is inappropriate. The strandlines are the main observational evidence for extensive bodies of water in the northern plains. Yet many of the linear features interpreted as strandlines may have other origins. The breaks in slope at the bases of ridges (Figure 3-15), for example, look remarkably like breaks in slope at the bases of ridges on the Moon (see Wilhelms, 1987, p. 111) that are formed by mass wasting. We clearly need more information. Better imaging and better altimetry, from which we could get a better idea of the continuity and horizontality of the "strandlines," would be especially useful.

Summary

Enormous floods occurred episodically throughout martian history. Many of the floods had discharge rates that ranged as high as 10^9 m^3 s^{-1}, 100 times the peak discharges of the largest known terrestrial floods and 10,000 times the average discharge rates of the largest terrestrial rivers. Many of the floods appear to have been caused by catastrophic release of groundwater trapped below a thick permafrost. Floods could have been triggered by a variety of causes, such as volcanic activity, faulting, and large impacts. Individual floods could have involved as much as 10^5 km^3 of water. Those around the Chryse basin alone have brought as much as 6×10^6 km^3 of water to the surface and probably considerably more. This lower limit is equivalent to 40 m spread over the whole planet. Groundwater may have also leaked quietly or catastrophically into the canyons to form lakes in which thick sedimentary sequences accumulated. The lakes, in turn, may have drained catastrophically to form additional floods. Large lakes must also have formed downstream of the floods, and there is considerable evidence in the low-lying northern plains of the former existence of such lakes. Estimates of the sizes of these lakes range from modest sizes, covering less than 1% of the planet's surface, to ocean-sized lakes, covering as much as a third of the planet's surface. The fate of the water in the terminal lakes is unclear. The water may have simply frozen in place and be present as ice deposits, or water may have somehow seeped back into the groundwater system to participate again in flood events.

4 VALLEY NETWORKS

The valley networks are the most common drainage feature on Mars and the feature most commonly cited as evidence for former warm and wet climatic conditions. Because of their resemblance to river valleys, they have generally been assumed to have formed by fluvial erosion. If formed by fluvial erosion, they imply warm, wet climatic conditions, for under present conditions, small streams would rapidly freeze and prevent development of a network. Since most of the valley networks are old, the former warm and wet conditions are proposed mainly for early in the planet's history. These interpretations are, however, almost certainly too simplistic and the view of a warm, wet early Mars on which fluvial activity was common, changing early to the cold planet we know today, is likely to be only partly true, if true at all. The resemblance of the valley networks to terrestrial valleys is not exact. River channels, for example, are not observed within the valleys, as would be expected if they were fluvial valleys and other seemingly non-fluvial characteristics must be reconciled with a fluvial origin. In addition, while most of the networks appear to be very ancient, some, cut in young units, must have formed late. Thus, there are reasons to question both a purely fluvial origin and an exclusively old age for the networks. We are on questionable grounds if we invoke warm, wet terrestrial climatic conditions to explain the martian valleys without understanding the reasons for the differences between the terrestrial and martian examples, and also if we invoke wet conditions for the early period of intense network development, while calling upon some other circumstances to explain the younger networks. In this chapter, we examine the characteristics, distribution, ages, and mode of origin of the valley networks and attempt to draw some implications with respect to climate.

Before beginning the discussion, it is useful to amplify upon a few terms. The distinction between fluvial valleys and fluvial channels was emphasized in the previous chapter. Fluvial erosion may occur and river valleys form as a consequence of surface runoff, in which precipitation is channeled into small streams that converge to produce larger and larger streams. In addition, river valleys form by groundwater sapping. In this case, the river is fed by a spring, and the valley grows by headward erosion as a result of undermining and sapping at the spring. Most terrestrial drainage systems form by a combination of the two processes, and erosion exclusively by runoff or sapping is rare. Normally some fraction of the precipitation is removed by runoff and some seeps into the ground to resurface elsewhere. For groundwater sapping to occur, the water table must be maintained by precipitation somewhere, although not necessarily close to the sapping site.

Mass wasting is a general term applied to the dislodgement and transport of rock and soil downslope. In mass wasting, the materials are not carried within another medium, such as wind or water, they simply move downslope because gravitational stresses have exceeded their yield strengths. Mass wasting can be sudden, as in landslides, or slow and barely perceptible. Slow mass wasting is commonly referred to as *creep*. Although water is not necessary for mass wasting, it greatly facilitates the process. However, while ratios of hundreds to thousands to one of water to eroded materials are common in fluvial processes, only modest fractions of water are generally involved in mass wasting. I have a somewhat unorthodox view that mass wasting may have played a prominent role in the formation of the martian valley networks, so that the following discussion may be biased in that direction.

Most of the valley networks are in the cratered uplands. They form open, branching networks in which tributaries converge downstream. They are therefore very different from the outflow channels, which rarely have tributaries. Individual valleys generally have rectangular or U-shaped cross sections, with flat floors and steep walls. While valleys wider than 10 km are present, they are rare; most of the branches are just a few kilometers across, as compared with a few tens to hundreds of kilometer widths for outflow channels. In the uplands, branches smaller than 1 km across are also rare. The smallest valleys tend to end abruptly in alcove-like terminations and do not continue to divide into ever smaller valleys as terrestrial valleys, formed by surface runoff normally do. Individual networks are far less complex than terrestrial valley systems. The vast ma-

jority have fewer than 50 branches. The branching is usually open, with large undissected areas between branches. The networks themselves are also spaced apart, leaving large areas of the uplands undissected. Figure 4-1 shows a typical upland scene with narrow, branching valleys winding between the craters, with some valleys being sharply defined and others indistinct.

Distribution

Figure 4-2 shows the distribution of networks between 47.5°N and 47.5°S. As has long been recognized, the networks are found mainly in the heavily cratered uplands, designated Noachian in

Figure 4-1. Typical upland scene at 13°S, 336°W. Craters are in all stages of preservation, with even large craters having suffered substantial degradation. Valleys are also in different stages of degradation and wind between the craters. The valleys have flat floors and steep walls, from their sources to their mouths, in contrast to most terrestrial river valleys. The scene is 280 km across (Viking Orbiter frame 618A28).

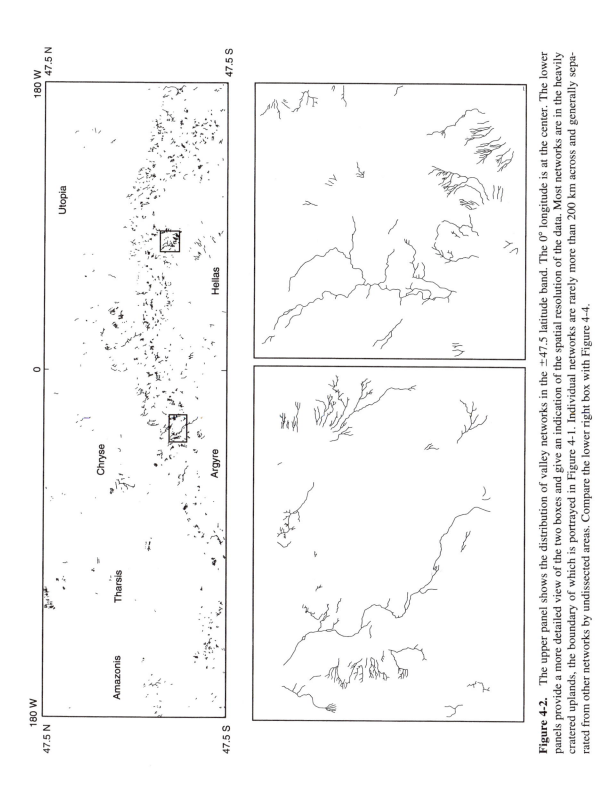

Figure 4-2. The upper panel shows the distribution of valley networks in the ±47.5 latitude band. The 0° longitude is at the center. The lower panels provide a more detailed view of the two boxes and give an indication of the spatial resolution of the data. Most networks are in the heavily cratered uplands, the boundary of which is portrayed in Figure 4-1. Individual networks are rarely more than 200 km across and generally separated from other networks by undissected areas. Compare the lower right box with Figure 4-4.

age, and dating from the period of heavy bombardment. Over 90% of the valleys shown in Figure 4-2 cut the Noachian. Most of the remainder are on volcanoes, or on steep slopes such as crater and canyon walls. Not all Noachian aged surfaces are heavily dissected. Arabia Terra, north of the equator, westward from 340°W to 10°W, is almost devoid of networks. Networks are also sparse in the Noachian of Arabia, north of 15°N, between 300 and 340°W, around the Hellas basin, and in Memnonia, between 150 and 180°W. Because most of the uplands are high, the networks are preferentially located at the higher surface elevations (Figure 4-3). Within the Noachian itself, networks are preferentially located at elevations of 2–5 km, reflecting the scarcity of networks in the low Noachian regions around Hellas and in western Arabia. Thus any proposed origin of the networks should account for both their preferential location in the Noachian and their preferential location in those parts of the Noachian at elevations greater than +1 km.

Even the regions of relatively high drainage densities are mostly poorly dissected. Local dissected areas are rarely more than 200 km across, and they are commonly 200–500 km from other dissected areas. The localization of the drainage within the dissected regions is the major cause of the speckled distribution pattern within the uplands seen in Figure 4-2. There appears to have been little or no competition between adjacent drainage basins, whereby one drainage basin enlarges itself at the expense of another (Carr and Clow, 1981). The pattern of dissection within the locally dissected areas varies greatly from open networks with few branches to dense networks in which the apparent drainage density is limited not by the networks themselves but by the resolution of the available images. Open networks are the most common (Figure 4-4).

Both local and regional patterns of dissection can be identified. Throughout the uplands are local highs, including rims of large craters, and local lows, such as crater floors and intercrater plains. Most of the local highs have a hilly or hummocky texture. Some of the lows appear as smooth plains or ridged plains that terminate abruptly against the surrounding upland. These plains are clearly younger than the upland. Where valleys are present in the adjacent uplands, they generally, but not always, terminate abruptly against these plains. But commonly the local lows merge imperceptibly with the local highs and channels that start on the highs fade away in the lows. Some networks converge upon and terminate within crater floors. Deposits are only rarely observed at the mouths of the trunk valleys.

The local nature of the drainage is illustrated

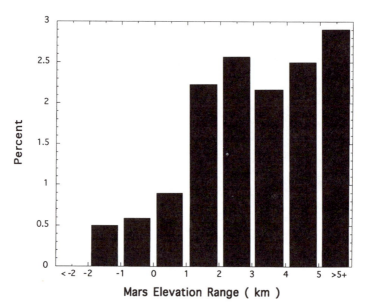

Figure 4-3. Distribution of valleys by elevation. The concentration at higher elevations is a result of their location mainly within the cratered uplands.

Figure 4-4. Networks at 20°S, 270°W, showing branching and rejoining to isolate islands. Large areas between network branches appear undissected, which results in drainage densities that are very low compared with most of the Earth. The scene is 280 km across (Viking Orbiter frame 625A27).

by Figure 4-5. The longest path through a network from the most distant tributary to the downstream outlet is the longest distance that eroded material must be transported. For almost 90% of the valleys, this distance is between 20 and 200 km; the average is 60–70 km and the longest is 1350 km (Carr, 1995). For reference, the Earth has 25 rivers over 1000 km in length; the longest, the Nile, is 6650 km long. The martian figures have probably been biased toward lower numbers by a variety of processes. Abrupt termination of downstream portions of many networks against smooth plains suggests that the lower reaches are buried and that the networks were formerly larger. In addition, the continuity of some networks is broken by superimposed craters, younger deposits, or simply by poor imaging. We can, however, conclude that any mechanism for valley formation must allow for erosion of valleys up to 1000 km long but must be such that in most cases development of the drainage system is arrested before the longest pathlength reached 200 km.

The local drainage patterns are superimposed on a regional drainage pattern (Figure 4-6). In some areas the regional pattern dominates; in

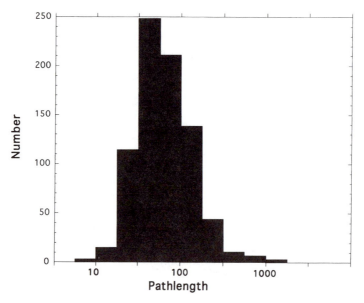

Figure 4-5. The number of channels plotted against the longest path (pathlength) through the network. Indicates the maximum distance that material must be transported to form the network. Over 90% of the networks have pathlengths less than 200 km.

others the local pattern dominates. In Arabia, from 300°W to 360°W, and along the plains/upland boundary from 180°W to 270°W, the drainage is mostly northward, down the regional slope to the plains/upland boundary. In the Noachis region, from roughly 0° to 40°S and 330°W to 20°W, drainage is mostly northwest, into the low-lying region of chaotic terrain east of the canyons and south of the Chryse basin. Within roughly 2000 km of the center of Hellas, drainage is mostly radial toward Hellas. The regional drainage patterns just described define four large areas in which drainage is toward Hellas, Utopia, Arcadia, and Acidalia. The areas are defined not just by the networks but also by the outflow and fretted channels, and clearly reflect variations in the global topography.

Network Topology

Although considerable efforts have been made to extract information on geomorphic processes from the topology of terrestrial river systems, the efforts have met with little success. Nevertheless, topologic techniques may be useful in quantifying differences and similarities between the martian networks and terrestrial drainage systems. The quantitative approach to fluvial network analysis was pioneered by Horton (1945) and Strahler (1958; 1964). In the Strahler method streams are ordered as follows. The smallest fingertip tributaries are designated first-order streams. Merging of two streams of equal order u forms a stream of order u + 1. Merging of streams of unequal order form a stream with an order the same as the highest order of the merging streams. The order of the network is the order of the highest order stream in the network, that is, the trunk stream. A network may additionally be thought of in terms of links (Shreve, 1966; Jarvis, 1977). Exterior links have a source upstream and a junction downstream; interior links have junctions upstream and downstream. First-order streams are single links; higher order streams generally consist of multiple links. Horton (1945) formulated two laws pertaining to stream networks. The law of stream numbers states that

$$N_u = R_b^{k-u},$$

where N_u is the number of streams of order u in a network of order k. R_b, the number ratio, is called the *bifurcation ratio*. The law of stream lengths states that

$$L_u = L_1 R_L^{u-1},$$

where L_u and L_1 are the average length of streams of order u and 1, respectively. R_L is known as the stream length ratio. For a large

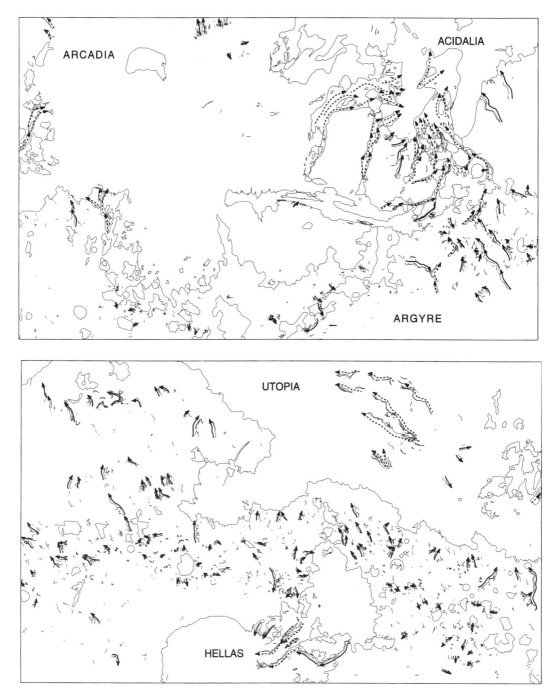

Figure 4-6. Directions of flow indicated by the geometry of outflow channels (dashed arrows), fretted channels, and valley networks (solid arrows). Also shown for reference is the Noachian-Hesperian boundary, the canyons, and the regions of chaotic terrain. (a) western hemisphere. Drainage is mostly toward Arcadia and Acidalia. (b) Eastern hemisphere. Drainage in the Noachian is mostly toward Hellas and northward toward the plains-upland boundary. Compare with Figure 3-1.

number of terrestrial river systems, R_b ranges from 3 to 5 and R_L ranges from 1.5 to 3.5 (Horton, 1945; Strahler, l964; Shreve, l966; Smart, l972).

The martian networks appear to follow similar laws (Carr, 1995). Figure 4-7 shows the number of branches as a function of order for 14 large networks. For these 14 networks, the bifurcation ratios range from 2.9 to 7.6, with an average of 4.3, values very similar to terrestrial river systems. The variation of length with stream order is less systematic, but the average length ratio of 2.9 is within the terrestrial range.

It is not clear whether these relationships have any value for determining what process are responsible for network formation. Shreve (1966) pointed out that Horton's laws are properties expected of networks with topologically random configurations. He examined all possible configurations in many different order networks and found that the networks with the highest probability of occurrence are those that follow Horton's law of stream numbers with bifurcation ratios typically found in terrestrial streams. Similarly, by assuming that link lengths vary randomly and are independent of location within the drainage basin, he demonstrated that networks

with random topology follow the law of stream lengths. The high probability of networks with stream length ratios of 1.5–3.5 follows because in topologically random networks streams of order u + 1 typically contain 1.5–3.5 times more links than streams of order u. Thus, any process that causes random development of a network should result in an array of networks in which the most common are those that obey Horton's laws. Despite several decades of effort and numerous claims, no clear relationships have emerged that relate topologic properties of terrestrial drainage networks to factors of geomorphic interest, such as basin relief, basin maturity, and climate (for a summary, see Abrahams, l984). The prospects of extracting information on process from the topology of martian networks are therefore bleak.

The martian networks have, however, one topologic feature rarely seen in terrestrial drainage networks, at least at comparable positions within the network. This relationship is shown in Figure 4-4 and in the lower right panel of Figure 4-2. At several places within the network, branches split and then rejoin to form islands. The largest island is 20 km across. Splitting and rejoining of terrestrial river channels to isolate islands is, of course, common, and the

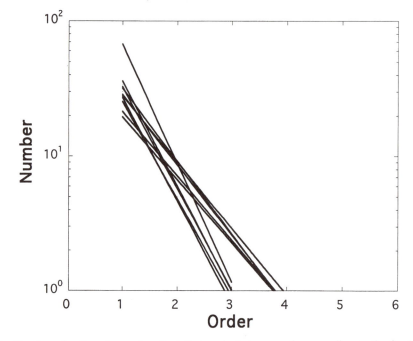

Figure 4-7. Number of valleys in a network plotted against stream order according to the Strahler ordering scheme. The relationships are similar to those for terrestrial river systems.

smaller islands in this image resemble islands within a channel. The larger island is more difficult to explain. Rerouting of a river to form areas completely surrounded by channels is common on a floodplain, where episodic floods may cause a river to repeatedly overflow its banks and cut new channels. But the area viewed in this image is clearly not a floodplain. Deeply incised valleys on Earth rarely, if ever, branch, with one branch striking across undissected country to rejoin the main valley tens of kilometers upstream. The simplest explanation of the relations seen here is that the channels (or valleys) extended themselves by headward erosion and the random headward extension ultimately led to intersection of two valleys or channels that had formerly branched. The relations shown in Figure 4-4 are not uncommon in martian networks. The rarity of such features in deeply incised, terrestrial valley or channel networks suggests that different processes may be involved in the headward erosion on the two planets.

Drainage Density

One network property that may be related to climate is drainage density. A number of workers have suggested that terrestrial drainage density (stream length per unit area) is related in some way to precipitation (Abrahams, 1984; Gregory, 1976; Gregory and Gardiner, 1975; Madduma Bandara, 1974). If such a relationship exists, it is complicated, being affected by the presence of vegetation, the maturity of the network, and the precipitation pattern. Terrestrial values for drainage density are mostly in the range of 2–30 km^{-1}. For the networks mapped in the more densely dissected parts of the martian uplands, the equivalent range is 0.001 to 0.01 km^{-1} (Carr, 1995). The large difference between these two numbers may be in part a reflection of the mapping scale. Nevertheless, the figures emphasize the vast difference (three to four orders of magnitude) in drainage density between the Earth and the martian upland. Clearly, whatever processes created the martian valleys, they were far less efficient than terrestrial fluvial processes or operated for only short periods of time.

Drainage densities on some volcanoes are significantly larger than in the cratered uplands. Compare, for example, Figures 4-1 and 4-2 with Figures 4-17 and 4-18 later in this chapter. Gulick and Baker (1990) estimated that drainage densities on parts of Alba Patera, Hecates Tholus, and Ceraunius Tholus ranged from 0.3–1.5 kin_1. These are the highest drainage densities measured on the planet.

Network Ages

The ages of the networks are difficult to determine. Generally, the small sizes of the valleys preclude reliable relative age determination by crater counting, although this has been attempted (Pieri, 1976). Superposition relations must be used to estimate ages. A channel superimposed on a geologic unit must be younger than, or contemporaneous with, that unit. Superposition, therefore, provides a maximum age. Despite the likely dominance of Noachian age networks, many networks are younger than Noachian. Scott and Dohm (1992) documented intersection relations for several hundred networks and suggested that possibly as many as 30% of them are Hesperian, or younger, in age. I determined superposition relations of 827 networks, using digital files of the networks and geologic maps, then checking the results by inspection (Carr, 1995). Out of the 827, 34 networks were assigned a maximum age of Amazonian, 34 were assigned to the Hesperian, and the remaining 759 to the Noachian. In terms of stream length, 146,000 km are cut into Noachian units, 8100 km into Hesperian units, and 10,300 km into Amazonian units. Differences between my results and those of Scott and Dohm (1992) are understandable. Ambiguities in age often result because of the difficulty of distinguishing between Noachian and post-Noachian units at the local scale. While extensive geologic deposits may confidently be assigned to one of the age groups on the basis of crater statistics, local deposits cannot. Valleys commonly meander between large craters in the uplands. The uplands as a whole and most of the large craters clearly date from the time of heavy bombardment, but younger deposits may have accumulated in some of the sparsely cratered areas between the large craters. Differences in the ages estimated for these intercrater surfaces lead to differences in the relative age assignments for the networks.

The abundance of networks in Noachian units

does not necessarily demonstrate that the networks are Noachian in age. However, the predominance of networks in the cratered uplands, the commonly observed abrupt terminations against local Hesperian deposits, the generally degraded appearance of networks when viewed at high resolution, and differences in dissection between local Noachian and Hesperian units suggest that most networks that cut the Noachian are indeed Noachian in age (Pieri, 1976, 1980a; Carr and Clow, 1981; Mars Channel Working Group, 1983; Baker and Partridge, 1986).

Despite the uncertainties, some general statements can be made about the ages of the networks. From 70% to 90% of the networks occur within the cratered uplands and are probably mostly Noachian in age. Most Hesperian networks occur where there is a substantial local relief, such as in the walls of Valles Marineris and along the plains-upland boundary. Extensive Hesperian plains such as Lunae Planum and Hesperia Planum are only very sparsely dissected. The youngest networks, the Amazonian, are found mostly on volcanoes where slopes are high and high thermal gradients are likely. Small networks that may be post-Noachian in age are also incised into the walls of some large craters such as Cerulli (Brackenridge, 1993). These age relations suggest that network formation was most efficient early in the planet's history, in the Noachian, but continued less efficiently into later epochs, particularly where abetted by high local slopes and/or high local heat flows, The valleys cut into Hesperian plains tend to be wider deeper, and have more open patterns then the networks cut into the uplands, as can be seen by comparing Figures 4-8 and 4-9 with figures of upland networks elsewhere in this chapter.

Networks—Channels or Valleys?

The importance of the distinction between channels and valleys was emphasized in the previous chapter. A river channel is the conduit in which a river flows; a river valley is normally the product of erosion of many streams and so contains many channels. A channel is comparable in width and depth to the river it contains. Since the late 1970s the dendritic networks have been widely referred to as valley networks (see, for example, Pieri,

1980; Mars Channel Working Group, 1983; Baker et al, 1992); formerly they were generally referred to as runoff channels (Sharp and Malin, 1975).

The supposition that the networks are fluvial valleys is based mostly on their planimetric form and on analogy with the Earth. But, despite frequent use of the term "valley network", there is little direct evidence that the networks are, indeed, formed of fluvial valleys although they clearly represent the result of surface drainage. A fluvial origin is supposed mainly because the planimetric patterns of branching sinuous valleys resemble the branching patterns of rivers on the Earth. Branches of most martian networks have flat floors and steep walls, and maintain a roughly constant width and depth over long distances. Widths are mostly a few kilometers and depths mostly 100-200 m (Goldspiel et al., 1993). Flat floors and steep walls are more typical characteristics of channels rather than fluvial valleys. Terrestrial river valleys may have flat floors downstream, but the upper reaches in the source regions are mostly V-shaped. Rarely do terrestrial valleys maintain a roughly rectangular cross section from mouth to source, as most martian networks do.

If the branches of the networks are fluvial valleys then they should contain fluvial channels. One example is a small channel in the floor of Al Qahira (419S11), but the inner channel extends only a small way up the main channel, not all the way, as would be the case if it were the channel of the river that cut Al Qahira. Despite some excellent high-resolution images of networks, we see no traces of river channels in high-resolution views of the flat floors of the network branches (see Figure 4-9 and many others frames listed in Goldspiel et al., 1993). The frequently observed branching and rejoining to form islands has already been noted as more characteristic of a fluvial channel than a fluvial valley. Ma'Adim Vallis, one of the largest of the networks has a prominent central ridge along most of its length. Long mid-channel ridges (Figure 4-10) are rarely, if ever, seen in fluvial valleys, and more resemble glacial median moraines than fluvial features. Levee-like ridges (Figure 4-11) occasionally seen in network branches, are never seen delineating terrestrial river valleys, although of course they are common on river channels. The ridges are particularly diagnostic because they probably indicate former fluid levels within the

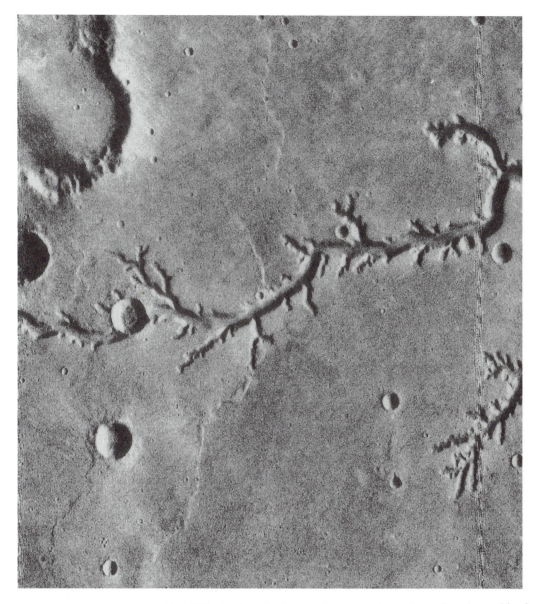

Figure 4-8. Nirgal Vallis at 27°S, 44°W, showing classic groundwater sapping patterns. Branches end in alcove-like terminations. Areas between the branches are undissected and the branching pattern is very open. The scene is 80 km across (Viking Orbiter frame 466A5).

channel. In summary, while the networks resemble terrestrial river valleys in plan and are widely interpreted as fluvial valleys, there are reasons for skepticism that they are indeed close analogs to terrestrial river valleys.

Fretted Channels

Fretted terrain is the term applied to a region along the plains-upland boundary between roughly 290-360°W and 30-50°N, in which the low-lying plains to the north complexly interfinger with the uplands to the south, thereby isolating numerous islands of upland surrounded by 1–2 km high escarpments (Sharp, 1973). Almost everywhere within the fretted terrain, debris aprons (Figure 4-12, with sharply defined flow fronts and convex-upward surfaces, extend from the escarpments across the adjacent plains for distances up to 20 km (Squyres, 1978). The debris

Figure 4-9. This high resolution view of a 2–km wide valley on the equator at 260°W is typical of valley networks in the uplands. The valleys have flat floors and steep walls and most lack smaller, narrower tributaries. Tributaries, if present, are commonly like the vague depression in the middle right of this image (Viking Orbiter frame 131S17).

aprons have been ascribed to mass wasting abetted by entrained ice. Of primary interest here is the origin of broad (up to 20 km wide), flat-floored, steep-walled channels, termed *fretted channels*, that reach deep into the upland (Figure 4-13), for they, like the debris aprons, may have formed largely as a consequence of mass wasting. The fretted channels branch upstream, as do

valley networks. They differ from the valley networks in being much wider and in displaying abundant evidence for mass wasting.

The debris aprons surrounding mesas and massifs, and on the valley walls, are the most recent products of erosion in the fretted terrain. Crater counts of Squyres (1978) indicate that those in Nilosyrtis date from roughly the base of the

Figure 4-10. Ma'Adim Vallis at 24°S, 182°W is one of the largest valley networks on the planet. A central ridge along part of its length resembles a median moraine. The size of the valley is difficult to reconcile with the modest drainage area and small number of tributaries. The scene is 250 km across (Viking Orbiter frame 597A20).

Amazonian. Others may be even younger. A major question is whether the processes that formed the debris aprons are also responsible for the gross dissection of the fretted terrain to produce the complicated pattern of mesas and channels. Where unconfined the debris aprons have clearly delineated outer margins roughly 20 km from the source. The aprons may be presently active and still moving, having reached only 20 km from their source because of the limited time since they started to flow. Alternatively, the aprons may be stable, having reached an equilibrium configuration. Irrespective of which alternative is true, the most recent episode of mass wasting represented by the debris aprons is unlikely to be responsible for formation of the fretted terrain. To form the fretted terrain, material must be transported for distances far greater than 20 km,

Figure 4-11. Levee-like ridges bordering a channel at 12°, 161°W. Median ridges are also present. The levee-likeridge may indicate the former level of fluid within the channel and provides supporting evidence for the supposition that the linear depression is some form of channel not a fluvial valley within which was a smaller river channel. The scene is 60 km across (Viking Orbiter frame 443S13).

and scarps must have retreated for distances far greater than the 5 km implied by the dimensions of the aprons (Squyres, 1978).

Many of the fretted channels have longitudinal striae on their floors. Some of the striae are clearly ridges. Where aprons from opposing walls meet, as in valleys, compressional ridges form (Squyres, 1978), and it is clear that some of the longitudinal ridges in the valleys formed sim-

ply by the meeting of opposing aprons. But not all the ridges and other longitudinal features have formed this way. Lucchitta (1984) points out cases where striae from the walls curl downstream and become longitudinal striae. She further shows examples where the longitudinal features in the center of the main valley transect striae from side valleys. These longitudinal features almost certainly formed by down-valley

Figure 4-12.   Massifs surrounded by lobate debris aprons at 46°N, 311°W. The debris flows extend roughly 20 km from the massifs, and are thought to be mixtures of ice and rock debris, both shed from the massifs (Viking Orbiter frame 338S31).

flow. The continuity of the striae from the walls with the longitudinal striae suggests that mass wasting has both removed material from the walls and carried it down the valley.

Mass wasting cannot explain all characteristics of the fretted terrain. Removal of subsurface materials followed by undermining and collapse also appears to have played a significant role in the development of fretted channels (Lucchitta,

1984). In Figure 4-14, several interconnected but partly enclosed depressions merge northward into a continuous fretted channel. In several sections, subsurface drainage has created a conduit into which the roof has collapsed, leaving fractured sections of the roof in a depression. The collapse features here are similar to those described in the previous chapter and attributed to collapse into an underground drainage channel

Figure 4-13. Typical fretted channel at 32°N, 341°W. Fretted channels are wider than typical valley networks and, unlike outflow channels, most branch upstream. They generally contain abundant evidence of mass wasting. The channel here is 8 km across for most of its length (Viking Orbiter frame 567A23).

from Ganges Chasma to Shabaltana Vallis. Completely enclosed channels also suggest subsurface erosion and/or solution, and the former movement of groundwater in the region.

Many of the fretted channels have attributes, such as alcove-like terminations of tributaries, U-shaped or rectangular-shaped cross sections, and no dissection of interfluves, which suggest headward erosion by sapping (Figure 4-15). However, none of the fretted channels show evidence of

fluvial erosion. They do not contain river channels, and the complex texture of their floors indicates clearly that the flat floors are not floodplains, although collapse of the surface in places to form linear depressions suggests subsurface flow.

One possibility is that the fretted terrain formed mostly early in the planet's history, mostly as a result of mass wasting along scarps, accompanied by water-lubricated creep of the re-

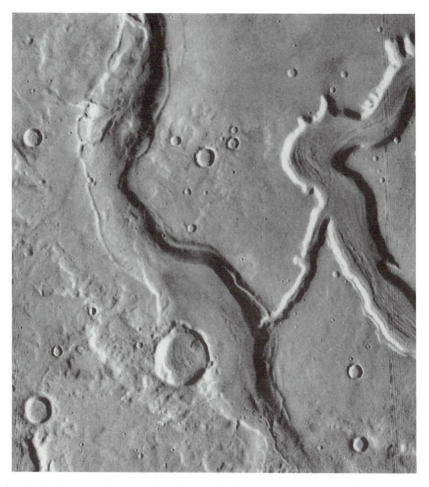

Figure 4-14. Detail of the top of the previous figure. The linear depression on the left appears to have formed by collapse, suggesting subsurface drainage. Compare with Figure 3-3. The texture on the floor of the 6-km wide channel on the right is typical of fretted channels (Viking Orbiter frame 205S16).

sulting debris (Figure 4-16), and subsurface erosion and solution (Carr, 1995). Preferential mass-wasting by sapping at valley heads may have created the valleys, the sapping process being aided by groundwater seepage into the mass-wasted debris. Flow of water across the surface may not necessarily have been involved; the debris could have flowed down the developing valley, away from the valley head, also by mass-wasting. Headward extension of the valleys would have been favored by the convergence of groundwater flow, and the consequent enhanced undermining of slopes at valley heads, loosely analogous to the way that terrestrial fluvial valleys extend themselves by groundwater sapping (Dunne, 1980). Then, as conditions changed, because of a declining heat flow, but also possibly because of

climate change, the 273 K isotherm may no longer have intersected the base of the debris flows, and the former liquid-lubricated flows would have stabilized (Figure 4-16). Talus subsequently shed from escarpments would have formed the viscous, markedly convex-upward, probably ice-lubricated flows that we currently observe, the ice being derived from ground ice in the uplands (Lucchitta, 1984). The transition from water lubricated to ice lubricated flows would have resulted in a major decline (to almost zero) in the rates of scarp retreat, planation, and headward erosion of the valleys. Formation of fretted terrains along the plains upland boundary in the 30-45°N latitude belt may have been favored by large local relief, drainage of groundwater northward toward the plains-upland boundary (Figure

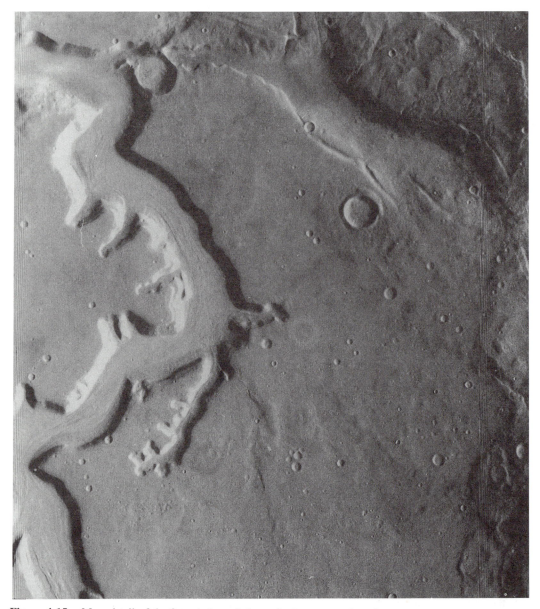

Figure 4-15. More detail of the fretted channel shown in the two previous figures. Here we see tributaries in various stages of development, each with the rounded, alcove-like termini that typify origin by sapping. Between the tributaries and away from the main channel, the surface shows no indication of drainage or any type of erosion. There are no fluvial channels within the main valley or its tributaries; mass-wasting features dominate. An issue here is whether the tributaries extended themselves by groundwater seepage that fed surface streams, or whether extension was simply by mass-wasting processes, possibly aided by groundwater flow, but without surface streams, as diagrammed in the next figure (Viking Orbiter frame 205S18).

4-6), and longer retention of ground ice (and groundwater at greater depths) near the surface at these high latitudes because of the lower loss rates by sublimation (Fanale et al., 1986).

Groundwater greatly facilitates creep and might therefore have enabled the eroded debris to travel long distances. Movement of fragmental debris is controlled by friction at the grain contacts. In a dry soil the compressive stress (mostly due to the weight of the overlying material) is distributed among the grain contacts, and for most materials as the compressive stress increases, the shear strength increases proportionately, the proportionality depending on the coef-

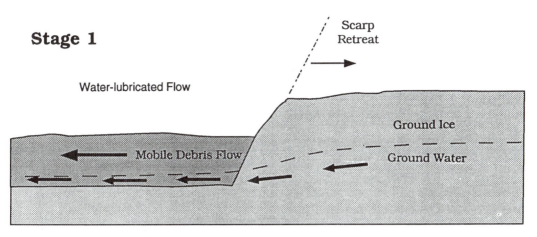

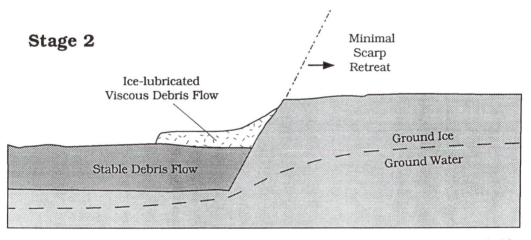

Figure 4-16. Diagram of a possible mechanism for formation of fretted channels. In stage 1, debris shed from slopes is mobilized by seepage of water into the debris, and headward erosion and scarp retreat are relatively efficient. In stage 2, water can no longer seep into the debris because of a change in heat flow or surface conditions, so that down-channel flow of the debris is inhibited, and ice-lubricated debris aprons develop.

ficient of friction of the material (Brunsden and Prior, 1984). With a pore fluid, part of the compressive stress is borne by the fluid and part by the solid skeleton. The compressive stress borne by the skelton is commonly called the *effective compressive stress* (Terzaghi, 1943). As pore pressure increases the fraction of the compressive stress borne by the fluid increases, and the compressive stress at the grain boundaries correspondingly decreases. Because the fluid has no shear strength and the resistance to movement at grain boundaries decreases with decreasing compressive stress, the effect of a pore fluid is to decrease the shear strength of the material. Ulti-

mately, as the pore pressure increases the material loses its shear strength and liquifies.

Groundwater leaking into the bases of debris flows in the fretted terrain may have so diminished their shear strength that the eroded materials could flow for long distances down the valleys. The presence of a permafrost zone could have enhanced the process by containing the groundwater, thereby enabling pore pressures to build, but the process may also work without a thick permafrost zone. Movement was likely slow, the result of jostling of the debris by temperature fluctuations, seismic shaking, and variations in the fluid pore pressures. When the temperature regime changed and the base of the debris flow froze, then the process of deformation and flow would have been very different. The shear strength of the flow would have been dependent of the combined shear strengths of ice and the rock skeleton, the combination depending on the configuration of the ice and rock materials. In this case, in contrast to the water case, the interstitial material (ice) has a finite shear strength, and so flow is inhibited, particularly on low slopes.

This discussion is very speculative and represents only one possibility. The presence of closed valley segments and roof collapse in the fretted terrain show that some form of subsurface erosion and/or solution was involved. Sharp (1973) speculated that dissolution of massive quantities of ice could also have contributed to formation of the terrain. The geometry of the terrain indicates massive dissection and removal of upland materials. Despite the speculation above, the exact nature of the processes, where the material was removed, and where it went remain mysteries.

Channels on Volcanoes

Most young channel networks are on volcanoes. They represent the strongest evidence against the simple climate change model in which early Mars was warm and wet but changed early to conditions that resemble the present and that persisted for most of Mars' history. The channels on the volcanoes are part of the reason that Baker et al. (1991) suggested episodic climate changes throughout Mars' history. If warm, wet conditions are invoked for early Mars to explain the channel networks, they should also be invoked to explain the relatively young channels on volcanoes, unless some channel-forming process unique to volcanoes has been operating.

The volcanoes with the most heavily dissected surface are on the relatively young, probably Amazonian, volcanoes Alba Patera, Ceraunius Tholus, and Hecates Tholus, although some channels are also on the older volcanoes Tyrrhena, Hadriaca, and Apollinaris Paterae (Gulick and Baker, 1990). The surfaces of Alba Patera and the Ceraunius and Hecates Tholii are among the most heavily dissected on the planet. They have closely spaced, subparallel, branching streams with small junction angles (Figures 4-17 and 4-18) that contrast markedly with the open upland networks with their winding streams and large junction angles.

The channels are also narrower and shallower, with widths generally under 0.5 km and depths of a few tens of meters (Goldspiel et al., 1993).

The origin of the networks on the volcanoes is even more uncertain than the origin of the upland networks because channels can form by purely volcanic processes, either as a result of erosion by lava or by fast-moving, hot, dense ash clouds, called *nuées ardentes* (Reimers and Komar, 1979). Many of the channels, however, branch upstream in a manner rarely seen with volcanic channels, and Gulick and Baker (1990), noting the similarity in the pattern of dissection of these and terrestrial volcanoes, concluded that many were probably valleys formed by surface runoff. They also noted (Gulick and Baker, 1993) that on terrestrial volcanoes the ratio of rock removed to water that passes through a channel may be as low as 1:1000, so that to erode the channels an abundant source of water is needed. One possibility is locally enhanced water erosion as a result of hydrothermal activity (Gulick and Baker, 1989, 1990, 1992). Mouginis-Mark et al. (1988) suggested that the channels on the flanks of Alba Patera formed by groundwater sapping in poorly consolidated ash deposits and presented a variety of arguments for the presence of ash.

Hydrothermal activity could have been important in the origin of networks, not only on volcanoes but also within the cratered uplands (Brackenridge et al. 1985; Gulick and Baker, 1993). As has already been noted, the geometry of most of the valley networks and the drainage density is inconsistent with surface runoff following precipitation as the primary cause of the valley networks. If the networks formed by fluvial action, then groundwater sapping is likely to have been the dominant process. Fluvial erosion requires that through the valleys must pass volumes of water that are hundreds to thousands of times the volumes of the valleys themselves. Brackenridge et al. (1985) and Gulick and Baker (1993) suggested that local heat sources could cause hydrothermal activity, which could both recycle local groundwater and draw upon more distant groundwater sources, and so provide the required volumes without precipitation. The heat sources could be volcanic or caused by large impacts. Water would rise over the heat source, seep onto the surface at springs, and be replaced by groundwater from afar or by water that seeped back from the surface into the local groundwater

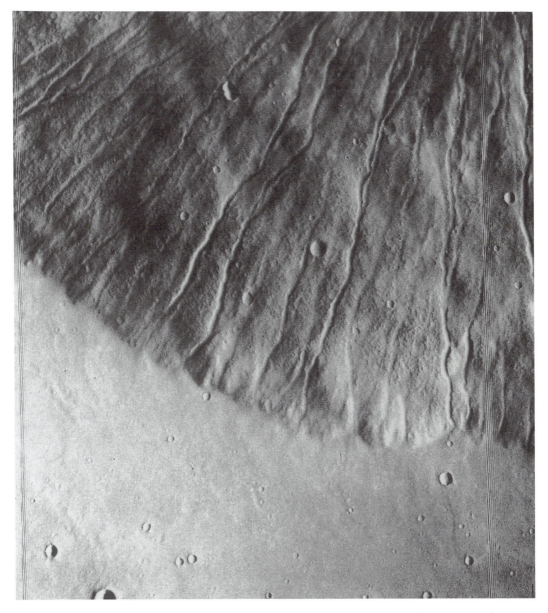

Figure 4-17. Channels on the flanks of the volcano Ceraunius Tholus at 24°N, 96°W. The channels are narrower and shallower than typical upland valleys, most here being under 100 m across. This volcano has one of the highest drainage densities on the planet (Viking Orbiter frame 662A56). Compare with the drainage density in Figure 4-4.

system if thermal conditions permitted. Modeling by Gulick and Baker (1993) indicated that the required volumes of water could be transported to the surface to cut the channels if magmatic intrusions reached the size of several hundreds of cubic kilometers and permeabilities were in the range of 10–100 Darcies.

Another possibility is that the flanks of the dissected volcanoes are comprised of ash, as suggested by Mouginis-Mark et al. (1988), and that the channels result from water-lubricated, mass-wasting of the ash. The presence of ice near the surface of the volcanoes is expected as a result from freezing of water outgassed from the erupted materials. The high heat flows on volcanoes would result in melting at shallow depths, lubrication of the poorly cohesive ash by liquid water, and mass wasting downslope. The shallow depth of the channels,

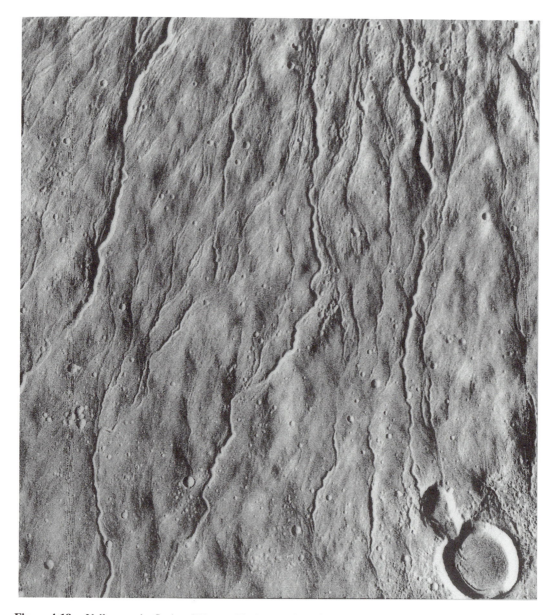

Figure 4-18. Valleys on the flanks of Hecates Tholus at 33°N, 210°W. As in the previous image, the valleys are mostly under 100 m across. The scene is 41 km across (Viking Orbiter frame 86A40).

typically a few tens of meters as compared with hundreds of meters for upland channels, would in this case be the result of the high heat flows. The water could be juvenile or, more likely, groundwater that has entered the volcano, thereby causing the pyroclastic activity and ash deposits.

Canyon Walls

Relatively young valleys are also incised into the walls of the canyons. They cut Hesperian plains adjacent to the canyon. Typically, the walls of the canyon are dissected into a spur-and-gully pattern, in which spurs of bedrock, reaching from the adjacent plateau down into the canyon, are separated by what seem to be talus chutes (Figure 4-19). In these sections, there is little need to invoke fluvial activity. In other sections, however, particularly between 77° and 87°W, deep valleys with V-shaped cross sections extend from the canyon floor deep into the adjacent plateau (Figure 4-20). They have alcove-like terminations,

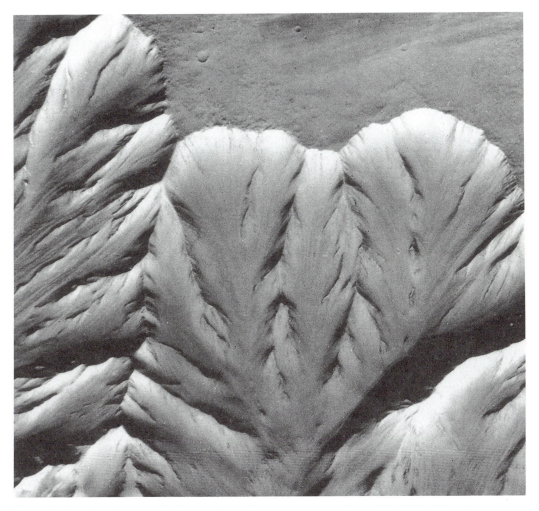

Figure 4-19. Typical spur and gully erosion on the steep canyon walls at 13°S, 69°W. Talus chutes are separated by jagged spurs of exposed bedrock. The image is 17 km across (Viking Orbiter frame 776A63).

like valleys formed by groundwater sapping, V-shaped cross sections, and steep walls. The areas between the large valleys are undissected, again suggesting sapping. One puzzling aspect of these valleys is the lack of deposits within the canyon, at their mouths, as large amounts of material must have been removed to form them.

Freezing of Streams

Attempts have been made to determine how fast a stream on Mars would freeze under different climatic conditions (Carr, 1983; Wallace and Sagan, 1979). The intent of the calculations was to determine if a small spring-fed stream could survive long enough to cut the observed valleys.

The techniques used are similar to those described for lakes in the previous chapter. The stream will form an ice cover. Temperatures at the surface of the ice can be calculated by balancing the heat losses at the surface, by sublimation and long-wave radiation, against heat gained by insolation and conduction through the ice from below. Heat lost by the water as a result of conduction into the walls and up through the ice is compensated for by freezing more ice onto the base of the ice cover. The results suggest that a 1 m deep stream flowing down a smooth conduit would take about 10 days to freeze under present climatic conditions. A stream 30 cm deep would freeze in a day. Calculations were done for different atmospheric pressures, taking into account

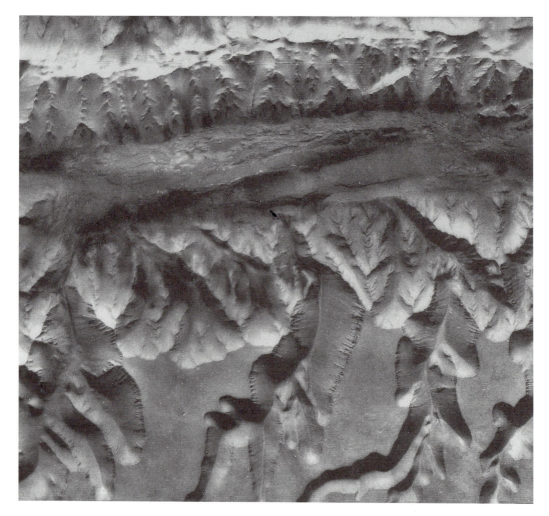

Figure 4-20. V-shaped valleys deeply incised into the south wall, Ius Chasma, at 7°S, 82°W. The far wall has the typical dissection shown in detail in the preceding figure. Between the deep valleys on the south wall the volcanic plateau is undissected. The adjacent plains are Upper Hesperian in age, so dissection to form the channels continued until at least this date and probably into the Amazonian. The mosaic is 110 km across (Viking Orbiter frame 66A08).

the increase in surface temperature and increases in conductive heat transfer to and from the atmosphere, but the results are insensitive to atmospheric pressure until pressures of several hundred millibars are reached. From the lifetime of the stream and the dimensions of the supposed channels, the distances flowed by the water can be determined from equations (3-1 through 3-5). For nominal slopes of 0.001, the distance a 1 m thick stream could flow, under these assumptions, is a few tens of kilometers.

While such calculations may be valid for smooth, obstruction-free conduits, they may have little relevance for natural situations. Spring-fed streams in sub-zero conditions on Earth typically form icings (Sloan et al., 1976; Williams and Smith, 1989). As the stream freezes, entrained ice tends to form dams. The hydraulic pressure builds under the ice and water bursts through the ice cover, redirecting the stream along a new path. The process continues to repeat itself. As a result, downstream flow is arrested and a broad ice sheet spreads across the valley. Presumably similar results would occur on Mars so that stream flow would be arrested long before the theoretical calculations predict.

This an important issue as it has been suggested that the valley networks could have formed exclusively by fluvial erosion of spring-fed streams under present climatic conditions (Squyres, 1989; Squyres and Kasting, 1994). There are several possible problems with this suggestion. First, discharges must be large to prevent freezing. Presumably as the discharge from springs is increased, a threshold would be reached at which icings would not form. One of the largest reported icings on Earth has a volume of 2×10^8 m^3 (Tolstikhin and Tolstikhin, 1974), implying a discharge rate for the spring of 10 m^3 s^{-1}, on the assumption that the icings start to form soon after average daily temperatures drop below 0°C. The threshold for formation of icings at this site is obviously larger than the present discharge, and the threshold must be much larger for martian conditions.

Squyres and Kasting (1994) acknowledged that, to perform the observed erosion under present conditions, the initial stream must emerge from the ground at least several meters deep. A stream several meters deep implies discharges of at least 10^3 m^3 s^{-1} for any plausible combination of slope and cross section (equation 3-3), so that these are very large streams. A second problem with sapping under present climatic conditions is that the large discharges implied by streams several meters deep would rapidly draw down the local water table and cut off supply. The spring would then freeze. Some mechanism must then be invoked to reactivate the spring at the same location after recharge so that sapping can continue and extend the valley. A third problem with sapping alone under present climatic conditions is that, because of the enormous volumes of water needed to cut fluvial valleys, some efficient mechanism for recharging the groundwater system is needed. There are thus good reasons to doubt that the networks could form by fluvial erosion under present climatic conditions. This topic is discussed further in Chapter 6.

Origin of Valley Networks

Since the early 1970s the networks have been widely accepted as having been cut by running water, and most of the discussions on origin have focused on the relative roles of sapping and runoff (Sharp and Malin, 1975; Laity and Malin, 1985; Pieri, 1976, 1980a,b; Baker, 1982, 1990; Carr, 1981) and the climatic implications of the valleys (Baker et al., 1991; Carr, 1983; Squyres, 1989; Squyres and Kasting, 1994). The fluvial hypothesis for valley formation has rarely been questioned. The pattern of dissection has been judged to be so similar to those of terrestrial rivers that a similar origin is implied. I have, however, recently proposed that mass wasting, aided by the presence of groundwater, may have played a dominant role in the formation of the valleys and that true fluvial erosion, that is, erosion by surface streams, may be rare, possibly being restricted to flood-like outbursts that could sustain themselves under present climatic conditions (Carr, 1995).

Groundwater sapping is the term applied to a process of valley formation in which a valley extends itself headward by undermining, collapse, and removal of rock materials at groundwater springs. Flow of water at the spring causes local convergence of groundwater and, thus, enhanced erosion and headward extension at the spring site. Terrestrial fluvial valleys formed by groundwater sapping tend to have abrupt, alcove-like terminations, U-shaped cross sections, undissected interfluves, and low drainage densities. Their configuration generally shows strong structural control, and the valleys tend to maintain their size for long distances downstream because of the small number of tributaries. These are attributes of the majority of the martian networks. Sapping characteristics are particularly well developed in some of the larger networks such as Nirgal Vallis and Nanedi Vallis (Figure 4-8). The evidence for sapping is so strong that some workers have suggested that the valleys formed exclusively by groundwater sapping (Pieri, 1976; Squyres, 1989) and that precipitation was not required.

Despite the clear dominance of sapping features, some networks have attributes more consistent with surface runoff. Formation of closely spaced fluvial valleys by sapping rarely occurs because the presence of one stream draws down the local water table, cutting off the supply to nearby areas. Yet some networks, particularly those on volcanoes, are so dense and intricate and so fill the drainage area (Figures 4-17 and 4-21) that a fluvial origin by groundwater sapping alone is unlikely. Another very common relationship, inconsistent with a fluvial erosion by groundwater sapping, is origin of channels at

Figure 4-21. Origin of valleys at crater rim crests at 22°S, 259°W. This relationship is unlikely with a ground-water sapping origin because of the improbable location of a series of springs at rim crests. High drainage densities in the center of the image and the lower right are also difficult to reconcile with a purely groundwater sapping origin. The scene is 280 km across (Viking Orbiter frame 625A29).

crater rim crests (Figure 4-21). Such an origin would require, improbably, that springs were located at rim crests where no local hydraulic head could develop, or require a combination of processes such as sapping, mass wasting, and outward erosion of the crater walls to capture the stream heads. Thus, while most martian networks have characteristics suggestive of sapping, any suggested origin must also account for the common inconsistencies.

Fluvial valleys, even if formed by sapping, are unlikely to be able to form under climatic conditions remotely resembling present conditions. As

discussed in the previous section, flow must be maintained once the water reaches the surface, and this is difficult under subzero conditions for small streams. Also, large amounts of water are required to cut fluvial valleys. Using sediment-to-water ratios found in terrestrial rivers, Goldspiel and Squyres (1991) estimated that the equivalent of several kilometers of water spread over the Ma'adim Vallis drainage basin were required to cut the Ma'Adim Vallis. Gulick and Baker (1993) suggested that much higher ratios of water to eroded materials are required to dissect coherent rocks. Erosion is the controlling factor not transportation, and ratios of 1,000:1 may be required. For a sapping origin some efficient recharge mechanism is clearly required. While recharge of the groundwater system at the equator by global movement of groundwater from the poles (Clifford, 1993) is possible, precipitation and seepage into the ground may be required to provide the quantities needed, and any such seepage would require significantly warmer climatic conditions.

I have questioned that the valleys formed, mainly by fluvial processes, on several grounds and have suggested that mass wasting might be the dominant process whereby material was eroded and carried down the valleys (Carr, 1995).

(1) Fluvial channels have not been identified on valley floors, despite excellent photography.

(2) Levee-like ridges on the sides some network branches suggest that some branches, at least, are channels of some kind, not fluvial valleys. They are unlikely to have been filled with water, for bedforms, like those in the outflow channels are not observed. More probably, what moved down the valley was the valley fill itself.

(3) The almost universal rectangular cross section and maintenance of almost uniform depths and widths from distant upstream tributaries to valley mouths are not typical attributes of fluvial valleys. The upper reaches of terrestrial river systems typically have valleys with V-shaped cross-sections, and these are rare in the martian highlands.

(4) The presence of longitudinal ridges in some valleys indicates that the floor is not a floodplain formed by a meandering stream, as expected if the valley is fluvial in origin.

(5) Many networks closely resemble fretted channels, particularly at high resolution, and for the fretted networks we have good evidence for mass wasting.

It was argued above that fretted channels may be true channels that formed mainly by mass wasting aided by the presence of groundwater, which abetted headward erosion and lubricated the resulting debris flows. While the evidence is less compelling, channel networks may have formed in a similar way. They have many of the attributes of fretted channels, except that most appear to be older and narrower, and they lack debris aprons. The resemblance between fretted

channels and networks is particularly striking along the plains-upland boundary between 180°W and 220°W, where several broad flat-floored valleys extend from the boundary deep into the uplands, the largest being the Ma'adim and Al Qahira Valles. Ma'Adim Vallis is 15–20 km wide and 1.5–1.8 km deep along most of its length. Its floor is mostly flat, with no trace of a central valley, although in places there is a central ridge, as in some fretted channels. Ma'adim Vallis has all the characteristics of a fretted channel, except that the flat floor is more heavily cratered and hence older than a typical fretted channel, and there are no debris aprons at the base of the walls. Ma'adim Vallis is also striking in that the large size of the trunk branch seems incommensurate with the modest size of the upstream drainage network. If formed by mass wasting, then the size of the upstream drainage basin is of secondary importance in determining the size of the channel in that most of the material that flowed down the valley was derived from the erosion that formed the valley itself and so was equivalent in volume to the valley itself. This contrasts markedly with fluvial erosion in which the volumes of water required to erode valleys are orders of magnitude larger than the volumes of material removed. Thus large catchment areas are generally needed to cut large fluvial valleys.

The resemblance between fretted channels and networks is also seen at high resolution. Numerous views of networks at resolutions of tens of meters per pixel (Figure 4-9, for example) show that most network branches, like fretted channels, have U-shaped or rectangular-shaped cross sections rather than V shapes, as expected of fluvial valleys (Goldspiel et al., 1993). This is the shape expected of channels that form as a result of plastic behavior (debris, ice, lava) rather than those that form by water erosion (Johnson, 1970; Chapter 14). Moreover, like fretted channels, the widths vary little along the path of the channel. Most exceptions to the U shape are where there are steep slopes as in the walls of the canyons and along some sections of the plains/upland boundary. In addition, some valley networks, like fretted channels, show evidence of subsurface erosion and/or solution, although this is rare.

The differences between terrestrial fluvial valleys and the martian networks, and the resemblance between the fretted channels and the networks led me to tentatively suggest that the networks may have formed by headward erosion, mainly as a result of mass wasting, and that the eroded debris was carried down the valleys not in surface streams of water but mainly by slow creep, possibly lubricated by ground ice and/or groundwater. Although mass wasting may have occurred on steep slopes, such as crater walls, without the aid of groundwater, ground failure at the stream heads and creep of debris down the channels would have been greatly facilitated by its presence. Channel formation could have been initiated at any steep slope. Once failure occurred, then further erosion and slope failure would have been favored at the scar because of the convergence of groundwater flow, as in the case of groundwater sapping on Earth (Dunne, 1990). The scar would develop into a channel that would continue to extend itself headward as long as seepage and the consequent slope failure were favored at the channel head and as long as the mass-wated material could move downstream. The process would have stopped on low slopes when climate change and/or lowering of the water table prevented further groundwater seepage into the eroded debris. The stream of poorly consolidated debris moving slowly down the valley would act as a high-permeability conduit for seepage of groundwater. The presence of groundwater at the base of the debris flow would lower the yield strength and facilitate movement of the debris. As suggested for fretted channels, creep of the eroded debris may have been facilitated by the presence of a permafrost zone that permitted high fluid pressures at the base of the debris flows. The process envisaged differs from fluvial erosion in two major respects. First, the movement of water is slow and mostly subsurface, being preferentially channeled within the poorly consolidated valley fill, because of its high permeability. Second, the valley fill moves down the valley en masse as a debris flow, lubricated by water-saturated debris at its base, not suspended or entrained in water.

A mass-wasting mechanism is consistent both with both open "groundwater sapping" patterns and with dense "surface runoff" patterns since a network would develop by headward erosion, perhaps aided by groundwater sapping, but the runoff is mostly the surface materials, not precipitation. Channels could also extend to crater rim

crests by dry mass-wasting. Levees are simply explained as marking former levels of debris within the channels. Median ridges are analogous to longitudinal moraines formed where streams of debris from different branches merge. Branching and rejoining of channels is explained by extension of channels headward by mass wasting.

While mass-wasting is envisaged as the dominant process, other processes may have contributed to formation of the channel networks. If groundwater was present, then subsurface erosion may have occasionally occurred, as with the fretted and outflow channels. Floods may also have occurred if artesian pressures were high enough, although there is little if any evidence for floods in the networks. What appears unlikely, in my view, is that the networks are comprised of river valleys that were cut mainly by slow erosion of running water in surface streams.

The main difficulty with the mass-wasting hypothesis is the mobility of the mass-wasted debris. Is fragmental debris sufficiently mobile that, given time, material could be transported over 1000 km? This question cannot be answered with assurance, and it is doubtful that modeling of the process will answer the question. But, as has been previously noted (Carr and Clow, 1981), whatever the process that created the channel networks, it was very inefficient. Dissected areas have drainage densities orders of magnitude less than the Earth's, and large areas are left undissected, despite a geologic record that spans billions of years. If we conservatively assume that the main active channel-forming period was very short, lasting only 100 Myr, from 3.8 to 3.7 billion years ago, then material must move 1 cm/yr to travel 1000 km. But we saw earlier that very few valleys are over 1000 km long; the pathlengths for almost 90% are under 200 km. Furthermore, the period of active channel formation is likely to have been much longer than 100 Myr. It appears likely, therefore, that movements of only millimeters per year or less are required to form most of the networks. As indicated under fretted channels, movement of mass-wasted debris would have been aided by jostling as a result of temperature fluctuations, seismic shaking, impacts, and variations in fluid pore pressure. The biggest uncertainty is the yield strength of the material.

The ages of dendritic channels suggest that their rate of formation declined rapidly after the end of the Noachian. Several factors, such as climate change, a declining heat flow, and loss of near-surface water, could have contributed to the decline. The valley networks occur mostly at low latitudes. Down-channel movement of debris appears to have continued to a later date in the fretted channels in the 30–50° latitude band. Persistence of movement to a later date at high latitudes suggests that perhaps progressive dehydration of the surface materials at low latitudes (Clifford and Hillel, 1983; Fanale et al., 1986) was a significant factor in the decline in channel development at low-latitudes. Loss of water from the low latitude uplands may have been caused by its instability and consequent sublimation, but slow migration of groundwater from high to low areas, following the pattern of surface flow shown in Figure 4-6, could also have been a contributing factor. The absence of debris flows and the crispness of the low-latitude terrains (Squyres and Carr, 1986) emphasize the efficacy of this loss of near-surface water at low latitudes and suggest that the process had already started in the late Noachian.

The discussion just completed is very speculative. The most widely held view of the origin of the valley networks is that they are fluvial in origin and formed by slow erosion of water in surface streams. Certainly, such an origin is supported by the planimetric pattern of dissection. The mass-wasting origin is not supported by any terrestrial analog, it being too slow a process to have left a significant mark on the Earth's surface. The fluvial origin is the consensus view. Despite a few contrary claims, such an origin almost certainly requires warm climatic conditions. A small fraction of the valleys are post-Noachian. If these are fluvial, warmer climatic conditions must have occurred occasionally late in martian history, as advocated by Baker et al. (1991). Another possibility is that the Noachian channels are fluvial and formed under warm conditions, but that the younger valleys, being mostly on volcanoes, or canyon and crater walls, where there were high heat flows and/or steep slopes, formed by mass wasting of ice-rich ground under climatic conditions similar to those that prevail at the present. Clearly, the origin of the valley networks and the climatic conditions required are both very uncertain.

5 GROUND ICE

In Chapter 2 we defined the martian cryosphere as the region of martian crust where if water were present it would be frozen. In the expectation that if groundwater was present it would have dissolved salts within it, a basal temperature of 252 K was assumed. The nominal thickness was estimated to be 2.3 km at the equator and 6.5 km at the pole, but these figures incorporate large errors because of poor knowledge of the thermal conductivity of the crust and the heat flow. At the end of heavy bombardment the heat flow is expected to have been four to five times larger than the present (see Figure 1-14), so the cryosphere should have been at least four to five times thinner than at present. But if the climate was significantly warmer at that time, the cryosphere may have been nonexistent or very thin. Thus, since the end of heavy bombardment the cryosphere at low latitudes has grown from an initial thickness of no more than a few hundred meters, and perhaps considerably less, to the present thickness of a few kilometers. We also saw in Chapter 2 that under present climatic conditions ice is unstable at low latitudes and that any ice present in the cryosphere at these latitudes at the end of heavy bombardment is likely to have been lost by sublimation down to depths of a few hundred meters unless it has, in some way, been replenished, or if climatic conditions were significantly different for much of the time. The arguments presented in Chapter 2 were almost entirely theoretical and were based on present surface conditions. In this chapter we examine the observational evidence for the presence of ground ice.

Terrestrial Permafrost

The permafrost zone on Earth is defined as that part of the ground where the temperature does not get above 273 K for two successive years. Regions of the Earth with a permafrost display a rich array of landforms unique to these regions (see, for example, Washburn, 1973, 1980; French, 1976). The thickest permafrost on Earth is in Siberia, where thicknesses in excess of 500 m are common and thicknesses up to 1400 m have been recorded. The maximum recorded thickness in North America is 610 m (Washburn, 1973). Overlying terrestrial permafrost is an active zone that experiences an annual freeze-thaw cycle, and the permafrost table acts as an impermeable barrier at the base of the active zone. Every fall a freezing front propagates downward from the surface, enclosing a thawed zone between the permafrost table below and the newly frozen ground above until all of the active zone is frozen. In spring, a thawing wave propagates down to the top of the permafrost.

Usually ice is not uniformly distributed in the permafrost but rather forms segregated bodies separated by ice-poor or ice-free zones. Ice may be present in various forms. Icings, or *aufeis*, are sheets of freshwater ice formed during freezing of a river, as discussed in Chapter 3. *Pingo ice is formed by freezing of lake ice.* Icings, pingo ice, and glacier ice may all be buried and incorporated into the permafrost. Ice may also be injected into the permafrost from below. The permafrost commonly contains *ice lenses,* which may be up to hundreds of meters across. They form during freezing as a result of migration of groundwater through the pores of the unfrozen soil or rock toward the freezing front. Near-vertical, V-shaped bodies of ice, called *ice wedges,* may also form by repeated fracturing as a consequence of contraction during winter and refilling of the fractures with water during spring. Ice lenses should be present in the martian cryosphere, because they form as a natural consequence of the freezing of saturated ground. Whether the other forms of ice are locally present will depend on the geologic and climatic history of the area.

The unique processes that occur in terrestrial

permafrost regions result in a number of geologic and geomorphic features not found in warmer climates. The following very brief discussion is intended merely to explain a few commonly used terms. Various kinds of patterned ground are characteristic of permafrost regions. _Polygonally cracked ground_ may result from contraction of the permafrost during winter, in a process somewhat analogous to the formation of desiccation cracks in lake sediments. The cracks may partly fill with water or sediment in early spring, thereby leading to compressive stresses within the polygons when the temperatures in the permafrost rise later in the year. The polygons may range in size up to about 100 m across. A similar process can lead to striped ground if the surface is sloping. Other forms of patterned ground are circles, polygons, and stripes formed of rocks, which may be sorted or nonsorted. They form by frost heaving in a manner only poorly understood. _Frost heaving_ is a process whereby coarser rocks are brought to the surface. It is thought to result from a combination of upward pulling, caused by expansion of the wet, finer grained, enclosing materials during freezing, and upward pushing as a result of freezing of water that collects in gaps below the rocks.

Mass wasting is particularly common in permafrost regions. _Frost creep_ is the "ratchet-like downslope movement of particles as the result of frost heaving of the ground and subsequent settling upon thawing, the heaving being predominantly normal to the slope and the settling more nearly vertical" (Washburn, 1967). _Gelifluction_ is the slow downslope movement of water-saturated debris. Such movement (also referred to by the more general term _solifluction_) is particularly common in permafrost regions because the permafrost table prevents seepage of rainwater and meltwater into the ground. The near-surface materials then become saturated and susceptible to flow. Movements of centimeters per year have been measured on slopes of a few degrees. _Rocks glaciers_ are glacier-like tongues of rock debris that also move at rates measured in centimeters per year. They may be ice cemented or ice cored, and grade into true glaciers as the proportion of ice increases. The precise mechanism whereby rock glaciers move is uncertain (Wahrhaftig and Cox, 1959).

Thermokarst is a general term applied to areas where ground ice has been removed as a result of climate change or disturbances in the local thermal regime. The stability of ground ice is sensitive to the annual cycle of ground temperatures. Changes in the albedo of the surface can result, for example, from erosion, fires, or changes in vegetation. The consequent dissolution of the ground ice results from a combination of downwearing and backwearing at slopes. Local hollows, called _alases_, may form in which lakes can accumulate. The depressions may merge to form continuous valleys. The general result is a poorly integrated drainage system with numerous lakes and closed depressions. Within some lakes _pingos_ form. These are ice-cored mounds that form in the final stages of the freezing of a lake, when water and sediments on the lake floor are forced upward through the overlying ice by the high artesian pressures that build as the lake progressively freezes.

Comparisons have been made between terrestrial cold-climate features and martian landforms (Carr and Schaber, 1977; Rossbacher and Judson, 1981; Lucchitta, 1981), but care must be taken in making such analogies because, first, the exact process whereby many terrestrial permafrost features form is not known and, second, many terrestrial permafrost features result in some way from freezing and thawing in the active zone at the surface. Present climatic conditions prevent this happening today on Mars, and it is very uncertain when, if ever, during the time period represented by the surface record it did happen. In addition, most of the terrestrial features just described are so small that they could not be detected in most of the available imagery.

Rheology of Ice

Before examining the morphologic evidence for ice on Mars, it is useful to examine briefly the rheologic behavior of ice as it provides clues as to how ice-rich soils might behave on Mars. The mode of deformation and the deformation rate of ice depend on its temperature and the shear stress. Shoji and Higashi (1978) constructed a deformation map of ice that shows where in the temperature-stress field the different deformation modes dominate (Figure 5-1). At low stresses creep occurs by lattice diffusion in which point defects flow through the bulk grains, and by

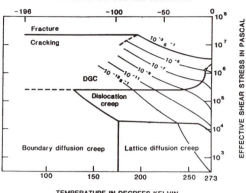

Figure 5-1. Plot of ice deformation processes as a function of temperature and the applied stress. DGC is dislocation glide with cracks. The cross curves with exponential notations indicate the strain rate. Stresses estimated for the base of the debris aprons are in the range of $1–5 \times 10^4$ Pa. (Adapted from Hoji and Higashi, 1978. Reproduced with permission from the International Glaciological Society.)

boundary diffusion, in which the defects flow through the grain boundaries. The strain rate varies with the applied stress, τ, and exponentially with $(-Q/kT)$, where Q is the activation energy for the deformation mechanism, T is the temperature, and k is Boltzmann's constant. Thus, at very low stresses the deformation is strongly dependent on temperature and is Newtonian, that is, the rate of deformation is proportional to the applied stress. At stress levels higher than about 10^4 Pa and temperatures above 170 K, dislocation creep, in which dislocations move, dominates and the behavior becomes non-Newtonian, with the deformation rate depending on τ^3 and the exponential term. At still higher stresses, deformation is by dislocation glide with cracks, and flow rate depends on the fifth power of τ. At very high stresses the ice fractures.

The main region of interest for Mars is the temperature range of 150–220 K. Because of the exponential term in the stress-strain relation, the strain rate drops dramatically with temperature. At effective shear stresses of 10^4 Pa (0.1 bars), those expected at the base of the debris aprons in the fretted terrain (Lucchitta, 1984), and temperatures of 205 K, the mean annual temperature at 40° latitude, the strain rate is approximately 10^{-16} s^{-1}. However, at a temperature of 165 K, the mean annual temperature at 70° latitude, the strain rate drops to around 10^{-20} s^{-1}.

Thus, if ground ice is present at latitudes higher than 40°, where it is thermodynamically stable, as discussed in Chapter 2, the effects of creep should be most evident in the 40–70° latitude band, where creep rates are highest. Moreover, if ice is present at low latitudes, where it is thermodynamically unstable, it should be very evident because of much higher creep rates.

The discussion thus far has been for pure ice. The rheology of ice rock mixtures is more difficult to assess and depends on the precise nature of the mixture. Small fractions of silicates would have little effect on the flow laws because the silicates would merely act as a suspension. However, when the silicate grains start to touch and friction at grain contacts becomes significant, then the rheology of the mixture will be affected. In addition, aggregates have flow properties that are different from continuous media. Shear stresses calculated assuming homogeneity underestimate the actual shear stresses because most of the shear is concentrated at grain boundaries. In ice-rock mixtures, additional complications may arise from segregation of the ice into lenses, as commonly occurs in terrestrial permafrost.

Debris Flows and Terrain Softening

Debris flows and terrain softening are perhaps the two most suggestive indicators that ground ice is present in parts of the martian surface. Both have been attributed to the slow creep of ice-containing materials, and both occur primarily in the 30–60° latitude bands, roughly where ground ice is expected to be stable and to have significant rates of creep (Squyres, 1979; Lucchitta, 1984; Squyres and Carr, 1986).

Lobate debris aprons

Lobate debris aprons were discussed in the previous chapter. They form at the base of escarpments and have convex-upward surfaces that slope gently away from the source escarpment, then steepen well away from the source to form a distinct flow front (Figure 5-2). On the surface may be striations that lead away from the escarpments in the assumed direction of flow. Transverse ridges may be also present, particularly where the flow encounters an obstacle. Transverse ridges also form where flows from opposing walls of a valley meet. In the northern hemisphere debris aprons occur wherever cratered

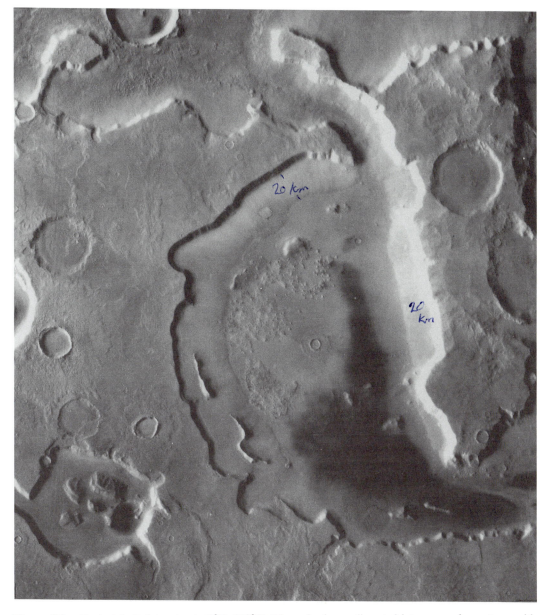

Figure 5-2. Typical fretted terrain at 40°N, 342°W. The uplands are dissected into a complex pattern, with steep escarpments separating upland remnants from lower lying plains. Debris flows extend about 20 km away from all the escarpments (Viking Orbiter frame 529A05). See also Figure 4-13.

uplands are found in the 30–60° latitude belt (Squyres, 1979). This is mainly in four longitude bands: in the Mareotis Fossae region (50–90°W), the Acheron Fossae region (130–140°W), in the Phlegra Montes (180–200°W), and Deuteronilus-Protonilus (280–360°W). In the southern hemisphere, they occur at the same latitudes, mainly around massifs on the rim of the Hellas and Argyre basins. The debris aprons appear to form wherever cliffs or steep slopes are present in the cratered uplands at the appropriate latitude. They are absent where similarly steep cliffs occur in the cratered uplands at low latitudes, such as along parts of the plains-upland boundary, or along the walls of outflow channels. Debris shed from these low-latitude slopes simply accumulates as talus. The debris aprons are also absent at very high latitudes, where extremely low creep rates are expected. Thus the lobate debris aprons are found roughly where ice is expected from its stability relations and where flow should be most evident because of the dependence of rheology on temperature. Attribution of the formation of debris aprons to entrained ice is thus very plausible, although not proven.

Two explanations have been suggested for the incorporation of ice into the lobate debris aprons in the 30–60° latitude belts. Squyres (l979) suggested that in these belts water ice became incorporated into the debris as a result of precipitation from the atmosphere. Frost was observed at the *Viking 2* landing sites. Its persistence for over 100 sols indicated that it could not be CO_2 frost and was likely to have been water ice. Squyres suggested that repeated precipitation of ice could result in incorporation of sufficient ice in the debris to affect its mobility. Lucchitta (l984) alternatively suggested that the uplands contain abundant ground ice and that it would be incorporated into any mass-wasted debris, thereby giving it the required mobility. The latter explanation appears more convincing, for it is difficult to see how, under present conditions, significant fractions of ice could become incorporated into the flows from precipitation. To significantly affect the rheology of the debris, at least a few tens of percent ice are probably required (Squyres, l979). The small amount of water precipitated from the atmosphere in winter would have to become rapidly mixed into the flow before it could sublime in summer, and such rapid turnover or burial by surface materials is unlikely. Moreover, in order to build up the ice content to a significant percentage, mixing into the flow must be effected without bringing ice already in the flow to the surface, where it would sublime. Incorporation of ground ice from the source seems much more plausible. The presence of ice in the surface material is also suggested by terrain softening.

Terrain softening

Terrain softening is a term applied to a characteristic style of terrain degradation in which surface features become rounded and subdued (Squyres and Carr, l986). Softened terrain is also most commonly found in the 30–60° latitude belts. Figure 5-3 is typical of the cratered uplands at low latitudes. Craters of all sizes, from the largest down to those almost at the limiting resolution, are crisply defined. The largest craters have flat, cratered floors; steep walls with blocky, angular

forms; and sharply delineated rim crests. In the intercrater regions, low ridges and escarpments have distinct, sharp edges. Superimposed on the whole scene is a population of small, well-defined, bowl-shaped craters. Figure 5-4 is an image of the cratered uplands at the same scale, but at 48°S, and here the scene is very different. The rim crests of most large craters are rounded and poorly defined. Although the age of this surface is probably similar to that in Figure 5-3, the small crater population is largely absent. The few small craters that are present are generally shallow. Most of the intercrater ridges are also rounded, although occasional sharp sinuous features, suggestive of flow fronts, are present. Figure 5-3 is typical of the cratered terrain at low latitudes. Figure 5-4 is typical of the so-called softened terrain in the 30–60° latitude belt.

Figure 5-5 shows additional characteristics of the softened terrain. The relatively fresh crater at the bottom of the picture still preserves its petal-like ejecta pattern, but the outer part of the ejecta is delineated by a rounded outward-facing escarpment rather than a rampart, as is normal at low latitudes. Within the craters is what has been termed *concentric crater fill*, characterized by roughly concentric ridges and grooves that typically outline lobes pointed down into the crater. Troughs around the flat floors, as in the upper crater, are not uncommon. More extreme forms of softening are seen in Figures 5-6 and 5-7. In Figure 5-7 mass wasting has filled lows between high-standing massifs, and the mass-wasted debris has clearly moved to form lobate flows. Thus the rounding process appears to merge with the process that caused the lobate debris aprons. The subjective impression of rounding of ridges is confirmed by photometric profiles across low-latitude and high-latitude craters (Squyres and Carr, l986). The low-latitude profiles have a jagged angular look, whereas the high-latitude profiles have a subdued rounded look. At latitudes higher than about 60°· the softening is less obvious than in the 30–60° latitude belts.

The softened features just described are most obvious in the cratered uplands, but similar features are also found on many of the high latitude plains. Concentric crater fill is common, and concentric ridges and grooves similar to those on crater walls are seen on some knobs and ridges within the plains. But the generally low relief of the plains, and the presence of other features

Figure 5-3. This scene at 16°S, 163°W is typical of the cratered upland at low latitude. Crater rim crests are sharply delineated; well-defined small craters are scattered throughout. The largest crater is 30 km across (Viking Orbiter frame 443S10). Compare with Figure 5-4.

such as polygonal fractures and various forms of patterning, makes softening on the plains less obvious than in the uplands.

Several objections have been raised against the interpretation that terrain softening is the result of ice-abetted creep. Zimbelman et al. (1988) suggested that the distinctive ridge-and-groove topography of the concentric crater fill is the result of erosion of layered sediments within the craters and that the general softening could be due to partial burial by eolian deposits. Parker et al. (1989, 1993) have suggested that linear features on ridges and isolated hills within the northern plains are strandlines marking water levels in former lakes rather than flow fronts caused by mass wasting. In addition, Clifford and Zimbelman (1988) argued that at the low temperatures that prevail at high latitudes, the creep rate of ice is several orders of magnitude lower than at low latitudes. Because the cratered uplands date back

Figure 5-4. This scene at 48°S, 340°W is typical of cratered uplands at latitudes above about 30°. The rims of many large craters are rounded and the small crater population is indistinct. The rounding or softening in this scene and the next three figures have been ascribed to creep of the near-surface materials because of the presence of ice. The largest crater is 37 km across (Viking Orbiter frame 575B59)

to the Noachian, when water or ice was abundant at low latitudes, as evidenced by the valley networks, why, they ask, did softening not occur at low latitudes at that time? Moreover, they argue that experimental evidence indicates that ice has a yield stress of 100 kPa and would need to be buried about a kilometer before it would flow.

I find these contrary arguments quite unconvincing. Burial by eolian debris would tend to produce smooth low areas with islands of the original, sharply defined, terrain poking through, not the generally softened terrain that we see.

While eolian erosion may have etched the surfaces of the concentric fill and debris aprons, the lobate, flowlike forms of the fill, coupled with abundant evidence of mass wasting provided by the lobate debris aprons and such glacier-like flows as seen in Figure 5-7, are convincing evidence that the surface materials have indeed flowed, despite theoretical arguments to the contrary. Experiments to determine yield strengths are generally conducted at much higher strain rates (10^{-7} to 10^{-8} s^{-1}) than envisaged here. While ice may have an effective yield strength of

Figure 5-5. Cratered upland at 33°N, 312°W. Roughly concentric fill within the craters suggests flow from the crater wall toward the center of the crater. Craters of all sizes have rounded rim crests. The scene is 60 km across (Viking Orbiter frame 196S11).

100 kPa for short-term experiments, it is likely to deform extremely slowly at low applied stress (Shoji and Higashi, 1978), as is implied by Figure 5-1. Very low creep rates can produce dramatic effects if sustained for billions of years.

Given that the surface materials have flowed, ice is the most likely cause for the dramatic difference between the equatorial latitudes and higher latitudes. Ice is known from terrestrial experience to enhance creep rates in mass-wasted

debris. The various channels and valleys are evidence of abundant water at or near the surface. Softening is absent where ice is thermodynamically unstable and present where it is stable. Softening is less evident at very high latitudes, where creep rates of ice are extremely low. The confluence of all these factors makes a compelling case for ice-abetted creep as the cause of the terrain softening.

The issue raised by Clifford and Zimbelman (1988) as to why we have no evidence of fossil

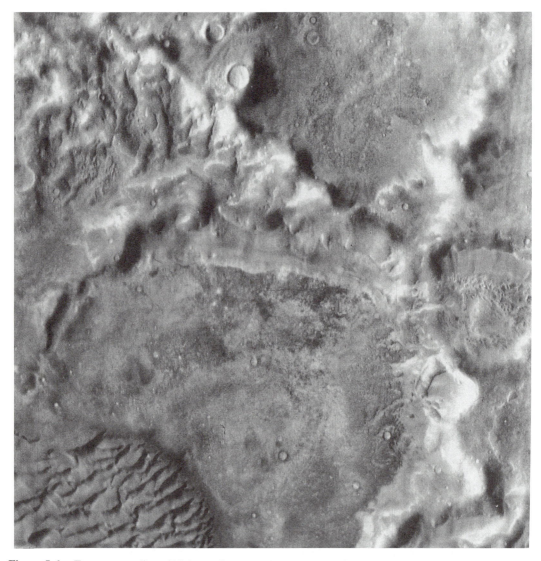

Figure 5-6. Extreme rounding of high-standing ground within the uplands at 48°S, 339°W. The closely spaced ridges in the bottom left corner are dunes. The scene is 56 km across (Viking Orbiter frame 575B59).

softening in the cratered uplands is a valid one. One possibility is that at the end of heavy bombardment the near-surface materials had already been desiccated to greater depths than those affected by mass wasting (Carr, 1986). Alternatively, the ground at both high and low latitudes was ice or water saturated at the end of heavy bombardment, the climate then changed, desiccation began, and the near-surface materials were depleted of ice before sufficient creep could occur to significantly alter the terrain.

Implicit in much of this discussion is an assumption that flow of near-surface materials caused the softening. Squyres (1989) has modelled the viscous relaxation of topography for different thicknesses of the viscous layer. For deep viscous layers (>5 km), long-wavelength components of the topography become suppressed, but the shorter wavelength components are preserved. Thus floors of large craters bow upward, but the sharp rim crests are preserved. This is not what we see in the softened terrain. To produce topography like we see in the softened terrain, the active, surface viscous layer must be some small fraction of a crater diameter in order to round the rim crests and preserve the flat floors.

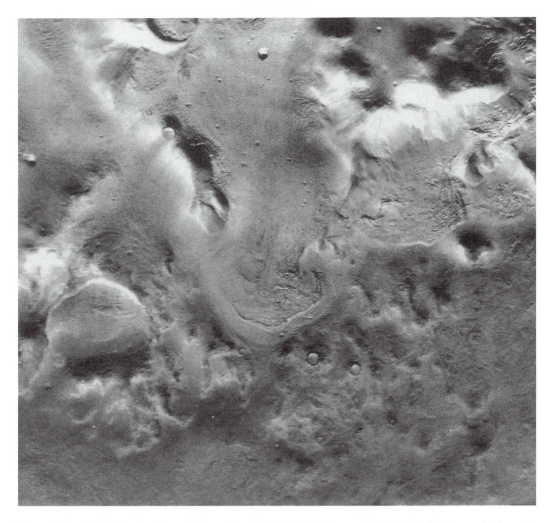

Figure 5-7. Debris flows and rounded massifs at 47°S, 247°W. The prominent, glacier-like tongue in the center of the image is 7 km across (Viking Orbiter frame 585B09).

Creep of material right at the surface is a reasonable possibility.

Rampart Craters

Martian craters have ejecta patterns distinctively different from those around impact craters on the Moon. Lunar impact craters have coarse, hummocky ejecta, close to the rim. Further out the texture becomes finer and grades imperceptibly into dense fields of secondary craters, and finally into discrete secondary craters and rays. This is true also of well-preserved martian craters with diameters larger than about 50 km. However, craters in the 2–50 km diameter range have very different ejecta patterns. The ejecta is commonly arranged in lobes, each lobe being outlined by a low ridge or rampart (Figure 5-8). The lobes wrap around pre-existing obstacles as though the material had flowed along the ground after being ejected from the crater. Flow along the ground is further suggested by the larger radial distances at which continuous ejecta are found around martian craters as compared with lunar craters of comparable size. Carr et al. (1977a) suggested, therefore, that, in contrast to the deposition of lunar ejecta, which is almost purely ballistic, the martian ejecta is emplaced by a combination of ballistic deposition and surface flow. After deposition, the ballistically deposited ejecta continues to move outward as a surface flow. They suggested further that the unique mobility of martian

Figure 5-8. Typical lobed ejecta patterns at 23°S, 79°W. The large crater in the center is 30 km across. Smaller craters, as in the bottom right and bottom left, tend to have a single annulus of ejecta, whereas the large craters have multiple lobes each outlined with a rampart (Viking Orbiter frame 608A45).

ejecta was the result of incorporation of ground ice or groundwater.

Since this original suggestion considerable work has been done on the variety of ejecta patterns and their distribution by latitude, elevation, and geologic unit. Experimental work has been done in attempts to reproduce the observed patterns, and theoretical modeling has led to a considerable improvement in our understanding of impact processes in general. However, despite the promise in the mid-1970s that ejecta studies

might dramatically improve our understanding of the distribution of volatiles at the martian surface, the results of many of these studies are somewhat confusing, even contradictory.

Schultz and Gault (1979, 1984) questioned the basic premise that the ejecta patterns are mainly the result of properties of the target. They conducted high-velocity impact experiments on dry fragmental targets, under different atmospheric pressures, and were able to reproduce many of the observed martian patterns. On the basis of these experiments, they suggested that the atmosphere could play an important role in creating fluidized depositional forms. They invoked a ballistic sorting mechanism whereby coarse material was emplaced ballistically, while the fine material was decelerated by the atmosphere and incorporated into a ground-hugging debris flow with entrained air, which ultimately formed the rampart pattern. They indicated that the target properties could affect the development of the flows by affecting the particle size distribution in the ejecta. Gault and Greeley (1978), on the other hand, have reproduced the lobate patterns with impacts into wet targets, albeit at low velocities.

With increasing size the lobate patterns of martian craters become more complex (Mouginis-Mark, 1979; Barlow, 1988). Craters smaller than about 10-20 km tend to have single lobes of ejecta that forms an annulus around the crater, outlined by a low ridge. Craters ranging from this diameter up to around 50 km in diameter have multiple lobes. Craters larger than 50 km in diameter tend to have ballistic ejecta patterns like lunar craters. The dependence of crater morphology on location is more ambiguous. Earlier claims (Johanson, 1979) of strong dependence of crater morphology on latitude have not been confirmed by subsequent studies based on thousands of craters (Mouginis-Mark and Cloutis, 1983; Barlow, 1988), nor does there appear to be a strong dependence on altitude (Mouginis-Mark, 1979). There is, however, one dependence on location that all observers appear to agree on. In the mottled, polygonally fractured plains, at the ends of outflow channels, impact craters tend to have an inner annulus of ejecta, marked not by a rampart but by an convex-upward front, much like lobate debris aprons. Outside this annulus are more typical ejecta, although ramparts are not always observed (Figure 5-9).

Craters smaller than a few kilometers in diameter do not have ramparts. If the formation of ramparts is due to the presence of ice or water in the ejecta, then the onset diameter for ramparts should provide a clue as to the depth where water is encountered. Kuzmin et al. (1988, 1989) made a global survey of the onset diameter for rampart ejecta patterns. They found a strong latitudinal dependence, as expected if the lobate patterns are due to incorporation of ice into the ejecta. At the equator onset diameters are in the 4–6 km range,

but they decrease poleward to 1–4 km at 65° latitude (Figure 5-10). Onset diameters are somewhat lower at high northern latitudes than at high southern latitudes. Kuzmin used impact models to estimate the depth to the ice-rich layer, on the assumption that the lobate patterns formed when the size of the transient cavity, derived from the crater diameter, was such as to penetrate the ice rich layer. He found that the depths to the top of the ice rich layer was typically 300–400 m at the equator, 200–250 m at 30° latitude, and around 100–200 m at 65° latitude, with somewhat higher numbers in the south. These numbers appear to be in conflict with the conclusion drawn from the lobate debris aprons and terrain softening that ice is present very close to the surface at latitudes higher than 30°. A possible explanation is that smaller amounts of ice are needed to cause significant creep than are needed to cause ramparts.

While the variation in onset diameters with latitude is an observational fact, the absolute values estimated for the top of the ice-rich layer should not be taken too literally. First, we are not absolutely sure that entrained volatiles are the cause of the lobate patterns. Second, we do not know how deeply into the ice rich layer the transient cavity must penetrate before the lobate patterns develop. Third, we do not know how rich in ice the layer must be to cause the lobate patterns. Fourth, the craters represent an integrated record over a long period of time during which the depth to the top of the putative ice layer may have significantly changed. Nevertheless, the results are encouraging in that they show the latitudinal dependence that is expected if the lobate ejecta does indeed result from ground ice.

Another technique for assessing ice content from crater ramparts is to measure the ratio of the diameter of ejecta ramparts to the diameter of the crater on the assumption that if more ice is incorporated into the ejecta it becomes more mobile. Several workers have found a strong dependence of this ratio on latitude, particularly in the northern hemisphere (Blasius and Cutts, 1980; Mouginis-Mark, 1979; Kuzmin et al. 1988; Costard, 1988). Squyres et al. (1992) show that for craters with diameters of 8–12 km, the ratio is 2.5 at the equator and increases to 3.2 at high northern latitudes, and to 2.8 at high southern latitudes. Again the data appear to indicate more ice-rich materials close to the surface in the northern plains,

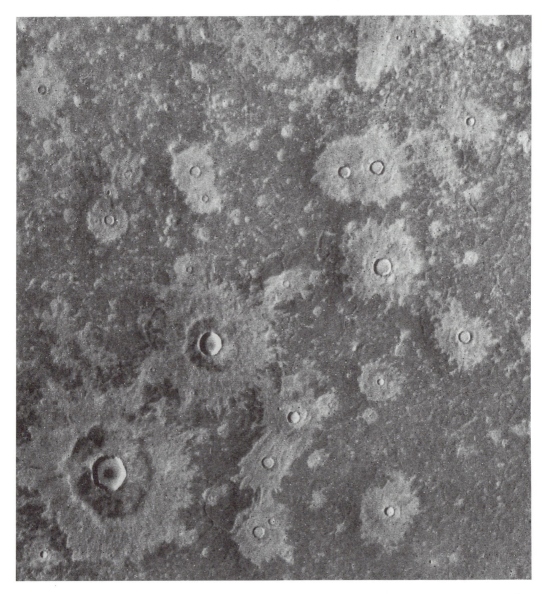

Figure 5-9. Mottled and polygonally fractured plains at 38°N, 258°W. The scene is 240 km across. These plains are situated at the end of large outflow channels in the low-lying northern plains. The distinctive ejecta pattern and the fracturing may be the result of the presence of ice-rich deposits. Compare also with Figure 3-14 which is similarly situated on plains near the termini of large channels (Viking Orbiter frame 673B30).

where ice deposits are suspected because of their location at the end of outflow channels.

Other Possible Indicators of Ground Ice

Polygonal fractures

We have already briefly discussed polygonally fractured ground in Chapter 3. Polygonally frac-

tured ground occurs mostly in the northern plains. When first seen, it was thought that the polygonal fracturing might have resulted in a manner analogous to terrestrial patterned ground in regions of permafrost. However, subsequent work (Pechmann, 1980; McGill, 1986; Lucchitta et al., 1986) has cast doubt that the terrestrial and martian examples are true analogs. The martian polygons generally range in size from 3 to 20 km

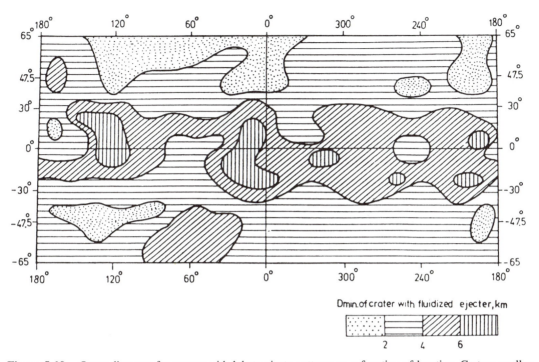

Figure 5-10. Onset diameter for craters with lobate ejecta patterns as a function of location. Craters smaller than the onset diameter do not have lobate patterns. Onset diameters are larger at the equator, suggesting the depth to ground ice is greatest at low latitudes. (From Squyres et al., 1992. Reproduced with permission from the University of Arizona Press.)

across, one to two orders of magnitude larger than typical terrestrial permafrost polygons. The troughs outlining the polygons are up a few hundred meters across and up to 100 m deep, and many of the depressions resemble graben in having flat floors. In general, the polygonal pattern is less regular than patterns typical of terrestrial permafrost. Polygonally fractured plains cover extensive areas of Acidalia Planitia and Utopia Plantia, at the end, respectively, of the circum-Chryse and Elysium outflow channels (Figure 5-11). Pechman (1980) argued that polygons of this size could not form simply by shrinkage as a result of annual cooling because the annual thermal wave penetrates only a short distance into the ground, and the polygons should have sizes comparable to the annual skin depth. Longer term temperature fluctuations, such as might be caused by obliquity changes, are unlikely to cause the polygons because the ice could creep to accommodate to such slow variations. He preferred a tectonic origin for the patterns. The occurrence of the polygonal ground at the end of outflow channels, and similar crater ages for the outflow channels and the patterned ground, suggested to McGill (1986) and Lucchittta et al. (1986) that the polygons are related in some way to the presence of ice-rich sediments. They suggest that the fracturing occurred as a result of cooling immediately after deposition, and/or self-compaction and accommodation of the sediments to the underlying terrain. They are therefore indicators of ice-rich ground in their view.

Patterned ground

Various forms of patterned ground occur in the northern plains. They include closely spaced stripes caused by variations in albedo, subparallel lines of pits or ridges, and lines that are combinations of pits and ridges. The subparallel, quasilinear features create a variety of patterns collectively termed *thumbprint texture* (Figure 5-12). They have variously been ascribed to strandlines marking shorelines of former lakes and oceans (see Chapter 3), terminal moraines of glaciers, and former positions of edges of eolian debris mantles, but their origins remain unclear.

Figure 5-11.   Polygonally fractured, low-lying plains at 41°N, 17°N. See also Figures 5-9 and 3-14. The picture is 58 km across (Viking Orbiter frame 32A25).

Fracturing and flow

Figure 5-13 is an example in which rigid slabs of volcanic material at the surface appear to have fractured and moved downslope on a mobile sub-surface layer. This example, at 34°N, 212°W, is at the northern edge of the volcano Hecates Tholus, but examples occur elsewhere. The simplest explanation is that the volcanic materials were deposited over an ice-rich substrate, which subsequently flowed as a result of warming of the ice following burial by the volcanics.

Thermokarst

Figure 5-14 shows an example of possible thermokarst at 23°N, 36°W. A plateau is outlined

Figure 5-12.   Moraine like features at 49°N, 284°W. The sinuous patttern is caused by sub-parallel ridges, lines of pits, and narrow depressions. These and similar patterns have been interpreted as caused by episodic retreat of a former shoreline, glacier, or depositional cover. The scene is 100 km across (Viking Orbiter frame 11B05).

by a jagged escarpment, similar to escarpments that form by a combination of backwearing and downwearing in terrestrial permafrost. Irregular hollows in the surface of the plateau are similar to terrestrial alases that form by removal of ground ice. The broad, irregularly outlined, low areas appear to have resulted from a merger of hollows in the plateau as a result of erosion and sublimation along the escarpments, much in the manner that valleys grow in terrestrial thermokarst. Similar merging alas-like hollows occur in Utopia Planitia at the end of the channels that start in Elysium (Costard and Kargel, 1995).

Layering in clearly visible on the walls of many of these hollow (Viking Orbiters frame 466B96).

Another possible form of thermokarst is shown in Figure 5-15. Here the ejecta of a 16 km diameter crater are outlined by an inward-facing escarpment. The ejecta appear to be inset into the surface. One possible explanation is that this area, being near the end of large outflow channels, is underlain by ice. Deposition of ejecta onto the surface disturbed the thermal regime and caused dissolution of the ice, in a way similar to the manner in which changing vegetation might trigger dissolution of ice in terrestrial permafrost.

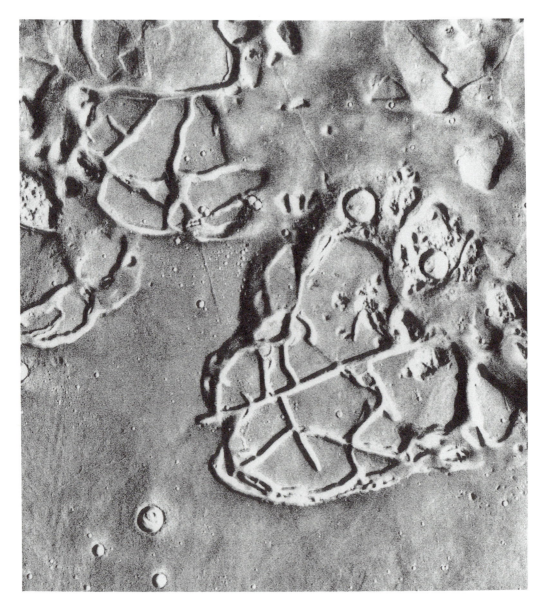

Figure 5-13. Mass wasting along the northern edge of the volcano Hecates Tholus, which is situated off the bottom of the picture. Volcanic materials at the surface appear to have slid to the north, down the regional slope. The simplest explanation is that movement of the rigid, near-surface, volcanic materials was facilitated by buried, ice-rich deposits. The picture is 42 km across (Viking Orbiter frame 86A35).

Miscellaneous

Other features have variously been interpreted as pingos, as pseudocraters that form by explosive eruptions when lava is deposited over ice-rich ground, and as table mountains, which are flat-topped, steep-walled volcanic edifices that form when a volcano grows through thick ice deposits (see below under volcano-ice interactions).

Glaciers

Glaciers are not truly ground ice, but it is useful to mention them in this context. The possibility that outflow channels could have been cut by glaciers, as proposed by Lucchitta (1982), was mentioned briefly in Chapter 3. Lucchitta pointed out the strong resemblance both in morphology and scale between terrestrial glacial features and

Figure 5-14.   Possible thermokarst at 23°N, 35°W. Irregular hollows have been etched out of the higher standing areas. Some of the hollows have merged to form valleys. The scene closely resembled terrestrial thermokarst caused by removal of ground ice. The scene is 45 km across (Viking Orbiter frame 8A74).

martian outflow channels. She envisaged spring fed glaciers, the springs creating massive icings that ultimately grew thick enough to move down pre-existing valleys and carve the large outflow channels. Recently a more radical suggestion has been made by Kargel and Strom (l992) that extensive glaciations have occurred relatively late in martian history. They draw analogies between a variety of martian landforms and terrestrial glacial features. Two landforms, which they describe as eskers and terminal moraines, are particularly intriguing. Eskers are narrow, sinuous, branching ridges of sand and gravel deposited from subglacial streams. They have the planimetric form of a stream channel but are positive relief features, left standing after the ice has disap-

peared. Esker-like ridges occur in several places on Mars, including near the south pole (Figure 5-16), within the Argyre basin, and adjacent to thick, easily erodible deposits along the plains-upland boundary, south of Elysium Planitia. The resemblance between ridges in these areas and terrestrial eskers is very striking. The putative terminal moraines are found mainly in Hellas and the northern plains and were mentioned earlier under patterned ground (see Figure 5-12). Kargel and Strom describe several other landforms in glacial terms but the case for a glacial interpretation is far less strong than with the "eskers" and "moraines."

It is vitally important for the climatic history of the planet to know whether these are indeed

Figure 5-15.   Detail from the mottled terrain shown in Figure 5-9. The ejecta around the large crater to the left appears to be inset into the surface. One possible explanation is that the area is underlain by ice-rich deposits and deposition of the ejecta disturbed the local thermal region, thereby causing ice below the ejecta to dissipate. Alternatively, the ejecta itself was ice-rich. After crater formation, material accumulated between the craters, then sublimation of the ice left the ejecta at a lower elevation than the surroundings. The image is 97 km across (Viking Orbiter frame 10B55).

glacial features. Kargel and Strom acknowledge that the formation of glaciers requires extensive cycling of water, so much so that oceans may be required to provide the precipitation. Moreover, the cycling must occur relatively late in martian history. Because of these radical implications, the uniqueness of the glacial explanation should

be examined closely. The "eskers" are found in places where there is evidence of removal of thick overlying deposits. This is true along the plains-upland boundary between 180 and 220°W, where there are numerous remnants of a thick, easily eroded deposit called the Medusae Fossae Formation; it is true at the south pole, where an-

Figure 5-16. Sinuous, branching ridges at 78°S, 40°W. The ridges resemble terrestrial eskers that form as a result of deposition from sub-glacial streams. These ridges may have formed under the nearby polar layered terrains at a time when they were more extensive than at present. Similar sinuous ridges are found in several other locations where there is evidence that a former sedimentary cover has been removed. The scene is 200 km across (Viking Orbiter frame 421B53).

cient layered deposits appear to have been stripped away, and it is true in the floor of Argyre, where an etched surface marks the remnants of former deposits. The sinuous branching pattern of the "eskers" is strongly suggestive of fluvial action. Observation of esker-like ridges emerging from beneath thick, highly eroded sedimentary deposits suggests that processes analogous to those that occur at the base of a glacier could have occurred at the base of these deposits. The deposits may be, or have been, mixtures of ice and silicate, as suggested by Schultz and Lutz (1988), and melting at the base of the deposits gave rise to basal streams as in a glacier. Other evidence for subsurface flow of water was men-

tioned in Chapters 3 and 4 (see Figures 3-4 and 4-14). Similarly, the moraine-like configuration of ridges in the northern plains could have nonglacial explanations. They appear to represent successive positions in the episodic retreat of some former cover, but they could be strandlines representing former shorelines of a lake (Parker et al., 1989, 1993), glacial moraines marking former positions of a glacier, or ridges representing former positions of a sedimentary cover. I earlier proposed the latter explanation (Carr, 1981) because of other evidence for former debris blankets on the northern plains.

My inclination is to be cautious of the glacial interpretation. While the esker-like and moraine-like features are very suggestive, other interpretations are possible. The glacial hypothesis requires major climate changes late in Mars' history, for which there is little other supporting evidence. On the contrary, the generally low cumulative amounts of erosion since the end of the Noachian (see Chapter 6) and the almost complete lack of dissection of Amazonian surfaces, except for a few volcanoes, argue strongly against any major climate excursions, accompanied by precipitation, late in martian history. In contrast, Baker et al. (1991) accepted the glacier interpretation and proposed that oceans formed episodically throughout martian history, thereby triggering short-lived but major climate changes.

Volcano-Ice Interactions

Since the end of the Noachian, the equivalent of a 0.4 km thickness of volcanics appears to have been extruded onto the martian surface, and possibly the equivalent of another 3 km may have been intruded into the near surface (Greeley and Schneid, 1991). If Mars does have abundant ground ice near the surface and groundwater at greater depths, then we should see evidence of interactions between the volcanic materials and ground ice and groundwater. Several kinds of interactions are to be expected. Surface flows or shallow intrusions might interact directly with any water or ice that might be present near the surface, or water might enter a magma chamber, thereby affecting the style of volcanism, or a magma chamber might act simply as a heat source at depth and create a hydrothermal system of warm circulating fluids.

A number of features on the martian surface have been proposed as analogs of some terrestrial constructs, such as table mountains and moberg deposits, that form when lava is intruded into or under ice (Allen, 1979; Hodges and Moore, 1979). Examples are common in Iceland (Allen, 1979) and Antarctica (Hamilton, 1972). Formation of a table mountain is a three-stage process. First, intrusions under the ice form pillow lavas on the floor of a sub-glacial lake formed by melting of the ice. Second, water enters the vent causing explosive activity and building an edifice of pillow lava fragments, tephra, and glass. Third, when the edific reaches the level of the lake's surface, less water gets into the vent and subaerial eruptions form flows at the surface. The result

is a flat-topped mountain of fragmental debris, surrounded by cliffs, and capped with lava flows that emanate from a summit crater. Moberg ridges form in a similar way except from fissure eruptions, and the last stage, capping by flows, is rarely reached. The result is long linear ridges with highly variable widths and irregular surface textures, the ridges being built of tephra, glass, and lava fragments. Allen (1979) intepreted numerous features in Elysium, Mare Acidalium, and Arcadia Planitia as table mountains and moberg ridges, thereby implying that glaciers had formerly been present in these areas. It should be pointed out, however, that many of the supposed table mountains resemble pedestal craters. These are impact craters whose ejecta is surrounded by a low outward facing cliff. Pedestal craters appear to form where the impact was in an easily erodible material, and the target material was subsequently removed except around the impact crater where the surface is protected from erosion by the ejecta (Arvidson, et al., 1976; Carr, 1981, p. 53).

The most unambiguous evidence of volcanism affecting water or ice at the surface is the initiation of large outflow channels near volcanoes (Figures 3-8 and 5-17) as discussed in Chapter 3. The volcanic activity could have triggered eruption of water onto the surface both as a consequence of melting ground ice and thinning of the permafrost seal so that groundwater could more readily access the surface. Baker et al. (1981) also invoked massive volcanism in Tharsis to trigger the large floods that they need for their episodic ocean hypothesis, as discussed in the next chapter. In addition, small fluvial-like channels emerge in places from under individual lava flows in Elysium (Mouginis-Mark, 1985), Hellas (Viking Orbiter frame 408S73), and Tharsis (Viking Orbiter frame 806A60). Finally, Squyres et el (1987) attributed many linear ridges and flow-like features along the plains-upland boundard between 190° and 220°W to a combination of moberg ridges and mobilization of the ice-rich upland materials as a result of intrusion of sills. They calculated that a 10 m thick flow on the surface would melt or vaporize the equivalent of a water layer about 3.5 m thick. In this case most of the heat from the flow is lost at the upper surface and does not participate in melting of the ground ice. Melting by a sill is much more efficient and the water column melted or vaporized

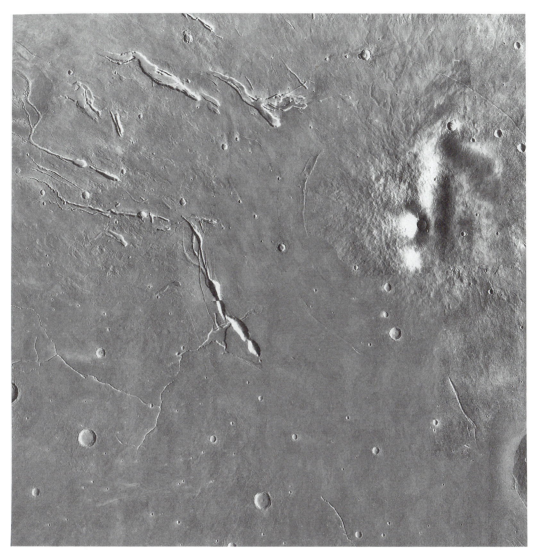

Figure 5-17. Channels starting around the periphery of the volcano Elysium Mons, whose summit crater, at an elevation of over 10 km above the datum, is in the right center of the image. Several rectilinear, graben-like depressions west of the volcano merge northwestward into sinuous, seemingly fluvial, channels. The image is 800 km across.

would substantially exceed the thickness of the sill.

Large amounts of groundwater circulating in the deep plumbing system of a volcano can result in explosive activity within basaltic volcanoes (Wilson and Head, 1994) and eruption of large volumes of ash. Tyrrhena Patera is surrounded by thick, extensive, layered, easily erodible materials that are most plausibly interpreted as ash deposits, the result of pyroclastic activity in the central volcano (Carr, 1981, p. 107; Greeley and Crown, 1990). Malin (1977) proposed that mut-

ing of surface features on Elysium Mons was caused partly by the presence of ash. Ash deposits have also been proposed for several other volcanoes, particularly those, such as Alba Patera and Ceraunius Tholus, that are dissected by channels (Reimers and Komar, 1979; Mouginis-Mark et al., 1988). On Earth, pyroclastic activity is more typical of andesitic volcanism than basaltic volcanism, but water entering the vents of a basaltic volcanoes can result in violent eruptions of ash as occurred, for example, at Kilauea in 1924 (Carr and Greeley, 1980, p. 171). We have no evidence of andesitic volcanism from the SNC meteorites, and the analyses of the soils at the Viking landing sites are consistent with a basaltic source, so intepretaiton of the pyroclastic deposits as the result of phreatomagmatic activity (reaction between magma and groundwater or seawater) is plausible.

Given an extensive groundwater system and sustained volcanic activity, hydrothermal activity must have occurred repeatedly throughout martian history. Heat sources may also have been provided by large impacts. As we saw in the previous chapter, hydrothermal activity has been invoked to explain valley networks on volcanoes (Gulick and Baker, 1990) and around impact craters (Brackenridge et al., 1985). In a typical hydrothermal system, a buried heat source warms groundwater which rises in a column over the heat source. When the column reaches the near-surface it spreads out laterally so that a mushroom shaped volume of warm groundwater is formed. The warm water rising from depth is replaced by cool water flowing in from the sides. Typical water circulation times are 10^4–10^5 years. Because of their biologic interest, hydrothermal systems are discussed further in Chapter 8.

Summary

A general muting of the terrain and presence of debris aprons at high latitudes suggests the presence of abundant ice at shallow depths. The absence of these features at low latitudes suggests that at the end of the Noachian the near-surface materials were ice poor or became so thereafter on time scales that are short compared with the time scales needed to modify the terrain significantly by creep. Onset diameters for rampart craters suggests ice-rich materials persisted at low latitudes much longer, and possibly to the present, at depths of a few to several hundred meters, but this conclusion is perhaps tainted by the possibility that rampart craters are the result of atmospheric effects rather than the effects of ground ice. In the low-lying northern plains, at the ends of outflow channels, a wide array of features, including polygonally fractured ground, unique ejecta patterns, possible thermokarst, and strong albedo contrasts between ejecta and local terrain, suggest the presence of thick ice deposits and/or ice-rich sediments. In view of Mars' volcanic history, we ought to see evidence of volcano-ice interactions if ice and groundwater are abundant near the surface. Outflow channels starting at volcanoes, fluvial features emerging from under lava flows, channels in volcano flanks, thick ash deposits on and around volcanoes, and features resembling terrestrial table mountains and moberg ridges have all been cited as examples of such interactions.

6 CLIMATE CHANGE

This chapter discusses two very different aspects of climate change. The first concerns changes induced by quasi-periodic variations in the orbital and rotational motions of the planet. These changes occur on relatively short time scales, have likely occurred throughout the history of the planet and so been superimposed on any long-term changes. Unless there are large amounts of exchangeable CO_2 sequestered in the polar layered terrains or the remnant caps, such periodic changes are likely to be small. Much of the discussion on quasi-periodic changes focuses not on observational evidence, but on hypothetical arguments as to what effects variations in the astronomical motions might have on various processes such as transport of dust and exchange of CO_2 between the regolith, atmosphere, and poles. While layering in sediments at the poles has been attributed to modulation of deposition and erosion by changes in surface conditions induced by planetary motions, the connection between the layered terrains and planetary motions has not been unequivocally established. In contrast, we have abundant observational evidence that Mars might have undergone long term, secular climate changes. We have already discussed at length some aspects of the evidence, the presence of valley networks, but erosion rates provide additional, possibly stronger, support for the supposition that conditions on Mars early in its history were very different from those that prevailed during most of its subsequent history. We will examine the timing and climatic implications of valley network formation and changes in erosion rates, then explore how the implied climatic conditions might have been attained.

Quasi-Periodic Climate Change

Changes in orbital and rotational motions

Periodic changes in the pattern and intensity of martian seasons can result from changes in the orbital and rotational motions of the planet. Eccentricity causes an asymmetry in the climatic regime of the two hemispheres. At present Mars is closer to the sun during southern summer, so southern summers are shorter and hotter than those in the north. In addition, dust storms commonly start during southern summer, some reaching global proportions. During northern summer the threshold for initiation of global dust storms is not reached, almost certainly because of the lower temperatures that result from the greater distance from the Sun. Dust storms affect atmospheric temperatures, circulation patterns, and transport of water. They may also affect the albedo of the seasonal and residual caps, thereby affecting their stability. They thus have broad climatic significance. The present eccentricity is 0.093, but during the present epoch it varies between about 0 and 0.13 (Ward, 1992). Short-term variations with a period of 10^5 years are superimposed on longer variations with a period of approximately 2 Myr (Figure 6-1). The range of 0–0.13 is approximate because extrapolations backward in time become very uncertain for times greater than 10^7 years. On time scales of this length or longer, the eccentricity may drift to higher values and becomes unpredictable. At low eccentricities climatic differences between the two hemispheres are small; at high eccentricities differences between the two hemispheres can be significant, as they are today.

A second motion that causes periodic changes in the seasonal patterns is precession, the slow conical motions of the spin axis of the planet and the normal to the orbit plane. The Mars spin axis precesses with a period of about 173,000 years and the axis to the orbit plane precesses with a period of roughly 70,000 years (Ward, 1973). These motions cause rotations in the line of equinoxes and the line of apsides (see Figure 1-1), both with respect to inertial space and to each other. The net effect is to cause an alternation of

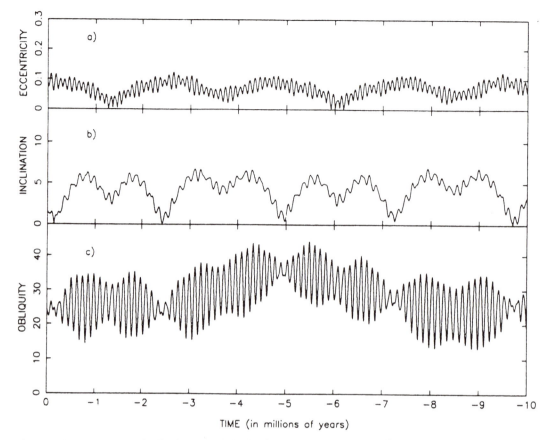

Figure 6-1. Eccentricity, inclination, and obliquity for the last 10 Myr, according to the orbit theory of Laskar (1988). Calculations vary significantly with very small changes in the initial conditions assumed. (From Ward, 1992. Reproduced with permission from the University of Arizona Press.)

the climatic regimes of the two hemispheres with a period of 51,000 years. Precession of the spin and orbital motions control the season at which perihelion occurs, and hence which hemisphere has the warmer, shorter summers.

Changes in obliquity are likely to have much larger climatic effects than the current variations in precession and eccentricity just described. The obliquity is the angle between the spin axis and the normal to the orbit plane. The present obliquity is 25.19°, but it undergoes large changes. During the current epoch, it is thought to oscillate between 13° and 42°, about a mean of 24° (Ward, 1992). The oscillations have a period of 1.2×10^5 years and an amplitude that is modulated on a 2 Myr cycle (Figure 6-1). There are considerable uncertainties as to what past obliquities were (Ward, 1992; Laskar and Robutel, 1993; Touma and Wisdom, 1993). Minute differences in the starting values for the calculations of

past motions lead to very large differences in the solutions when projected backward (or forward) in time, such that projections larger than 10 Myr have little meaning.

Obliquity variations are inherently chaotic on time scales of 10 Myr or longer. Part of the problem concerns resonances. If the period of precession of the spin is commensurate with one of the periods of variation of the orbit, then spin orbit resonances can occur. Excursions in obliquity significantly larger than are suspected from the current oscillations are then possible. Thus, secular changes in the orbital parameters can cause major changes in the precessional and hence the obliquity motions. All these variations cause the obliquity to be chaotic, at least on time scales greater than 10^7 years. Over geologic time values for the obliquity may have occasionally reached as low as 0° and as high as 60° (Laskar and Robutel, 1993; Touma and Wisdom, 1993). In this respect Mars differs from the other terrestrial planets. The obliquities of Mercury and Venus have been stabilized by tidal dissipation, and that of the Earth by the presence of the Moon.

Long-term changes can also be induced by the planet itself. Precession of the spin axis, for example, may change with time as a result of geophysical changes. The precessional period is given by $2\pi/\alpha \cos\theta$, where θ is the obliquity. The precession constant α given by

$$\alpha = \frac{3}{2}\left(\frac{GM_s}{\omega a^3}\right)\left(\frac{J_2}{\lambda}\right),$$

where G is the gravitational constant, M_s is the mass of the Sun, ω is the spin rate, a is the semimajor axis, $J_2 = [C - 1/2 (A + B)]/Mr^2$, and $\lambda = C/Mr^2$. A, B, and C are the principal moments of inertia, and r is the radius of the planet. Differentiation of the planet, or the building of volcanic constructs such as the Tharsis and Elysium bulges, will have affected the moment of inertia and hence the precession. As the precession changes, so do the resonances with other motions.

Obliquity affects the distribution of solar insolation with latitude (Figure 6-2). At low obliquities, the amount of insolation received, averaged over the entire year, depends strongly on latitude, with the poles receiving almost none. As the obliquity increases, somewhat less insolation falls on low latitudes, but the amount received by high latitudes increases substantially. At an obliquity of about 50°, the annual average is almost independent of latitude. At obliquities higher than 54°, more insolation falls on the poles than on the equator. Because the poles act as storage reservoirs for volatiles at medium to low obliquities, the onset of high obliquities has the potential for injecting large amounts of volatiles, particularly H_2O and CO_2, into the atmosphere.

Potential effects of changes in orbital and rotational motions

Changes in obliquity will affect how exchangeable CO_2 and H_2O is distributed between three main reservoirs: the atmosphere, the poles, and the high-latitude regolith. Of these reservoirs, only the volatile contents of the atmosphere and seasonal caps are well known, the atmosphere from direct measurements and the seasonal caps from pressure variations in the atmosphere (Table 6-1). The residual caps are bright areas

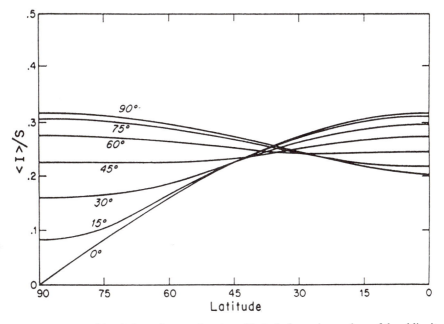

Figure 6-2. Average annual insolation <I> as a function of latitude for various values of the obliquity. The insolation is normalized to the average solar flux. At obliquities higher than 54° the average annual insolation is higher at the poles than at the equator. (From Ward, 1992. Reproduced with permission from the University of Arizona Press.)

Table 6-1. Exchangeable Volatile Reservoirs
(Modified from Kieffer and Zent, 1992)

	CO_2		H_2O	
	kg m^{-2}	mbar	kg m^{-2}	m^{a}
Atmosphere	150	5.6	0.01	10^{-5}
Seasonal cap	40	1.5	0.01	$<10^{-5}$
Perennial cap (N)	0	0	5.8	6×10^{-3}
Perennial cap (S)	8	0.3	600	0.6
Polar layered deposits[b]	0	0	6200	6.2
Polar layered deposits[c]	0	0	29000	29
Regolith[d]	1000	37.3	10	10^{-2}

[a] Equivalent depth spread over entire martian surface.

[b] Assumes layered deposits 1 km thick, 50% ice, and area as in Tanaka et al. (1988)

[c] Assumes layered deposits in the south 1.5 km thick, those in the north 5 km thick
(Dzurisin and Blasius, 1975), 50% ice, and areas as in Tanaka et al. (1988).
This is a probably an upper limit.

[d] This estimate is very uncertain. In the unlikely event that the regolith averages
several hundred meters of materials such as clays, with high adsorptive capacity,
then there may be as much as 250 mbar of CO_2 and several meters of water in
the regolith. The estimate also ignores the possibility of interstitial ice and
segregated ice bodies, particularly at high latitudes (Chapter 5).

that remain at the poles after retreat of the sea-sonal caps. The northern residual cap has an area of 837 km^2 (Tanaka et al., 1988), is estimated to be 1 km thick (Kieffer and Zent, 1992), and is thought to have water ice at the surface (Kieffer et al., 1976; Farmer et al., 1976). The southern residual cap, with an area of 88 km^2, is composed of CO_2. Its thickness is not known but it is likely to be between 1 and 100 m thick (Kieffer and Zent, 1992). If so, then it contains 0.6–60% as much CO_2 as the atmosphere. Thus, the residual CO_2 cap on its own has only a modest ability to buffer the atmosphere. If it were the only ex-changeable reservoir of CO_2, heating of the poles at high obliquity could increase the mean atmos-pheric pressure from its present mean of 5.6 mbar to only about 9 mbar. But two additional

sources of CO_2 have been suggested, clathrate within the polar layered terrains and CO_2 ad-sorbed on the regolith.

CO_2–H_2O clathrate is a spongelike structure in which H_2O molecules surround CO_2 molecules, in the ratio of 5.75:1, respectively. The clathrate is stable at polar temperatures at pressures in ex-cess of 100 mbar, or at depths greater than about 10 m (Miller and Smythe, 1970). If CO_2 and H_2O were incorporated into a polar cap and buried to depths greater than this, then clathrates could form. Jakosky et al. (1995) estimated that if the northern polar layered deposits were comprised largely of clathrates, then they could contain 0.2 bars of CO_2. If they are comprised of solid CO_2, they would contain 0.85 bars. Jakosky et al. sug-gested that much of the CO_2 from an early, thick atmosphere might be stored at the poles in this way. If so, then high obliquities could result in release of a several tenths of a bar of CO_2, along with considerable amounts of H_2O. Release of a several tenths of a bar of CO_2 would, however, raise surface temperatures only about 10–20 K, so that temperatures would remain far short of freezing. The suggestion of clathrates or large amounts of solid CO_2 in the polar deposits is somewhat speculative; there is no supporting observational evidence.

Exchange with the regolith also has the potential for effecting significant changes to the atmosphere by obliquity changes. The potential for obliquity changes to pump CO_2 in and out of the regolith and so affect atmospheric pressures depends on a variety of poorly known factors. The mechanism for exchange is adsorption and desorption from the regolith components in response to temperature changes. Temperature fluctuations

in the regolith can be estimated with some confidence. Recalling from Chapter 1 that the skin depth is $(\kappa t/\pi)^{1/2}$ and taking a value of 15 m^2 yr^{-1} for κ, the diffusivity, a value that is typical of frozen soils, we find that the skin depth is close to 1 km for 10^5 year obliquity oscillations. Obliquity induced temperature changes within the regolith are largest at high latitudes and at shallow depths (Figure 6-3). The main problems in estimating the obliquity effects are lack of knowledge of the structure and composition of the regolith, and the total amount of exchangeable CO_2.

The adsorptive capacity of rock materials depends on their mineralogy and grain size. Fanale and Cannon (1979) determined the adsorptive capacity of ground basalt and the clay mineral, nontronite, at martian temperatures and pressures (Figure 6-4). Nontronite, being a clay, has a large surface area and a significantly larger adsorptive capacity than ground basalt. The adsorptive capacity is temperature sensitive. Lowering the temperature of nontronite from 230 K to 150 K

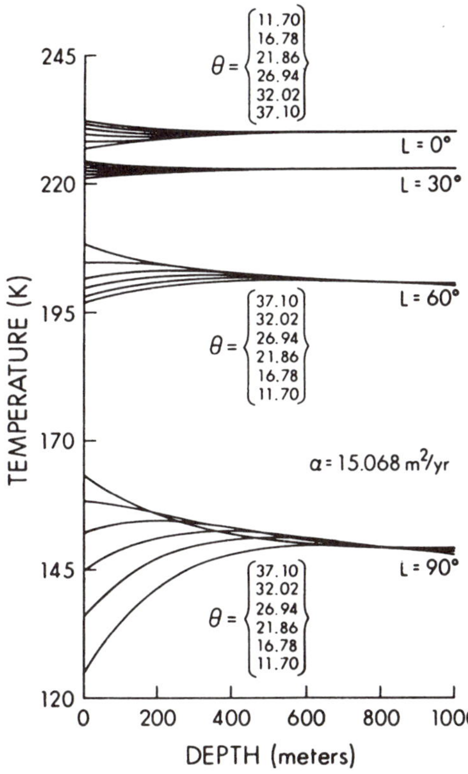

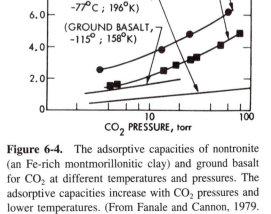

Figure 6-3. Temperatures in the regolith at four latitudes, L, during periods of increasing obliquity, θ. Internal heat flow is neglected. As temperatures rise at high latitudes, adsorbed CO_2 and H_2O will tend to be driven out of the regolith. (From Fanale et al., 1982. Reproduced with permission from Academic Press.)

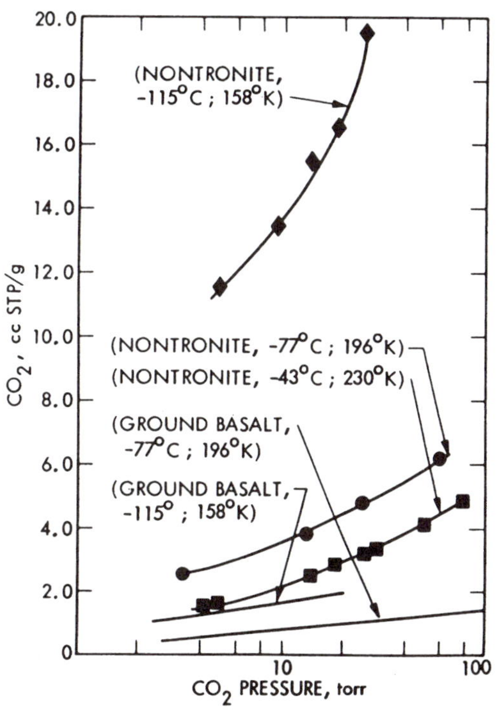

Figure 6-4. The adsorptive capacities of nontronite (an Fe-rich montmorillonitic clay) and ground basalt for CO_2 at different temperatures and pressures. The adsorptive capacities increase with CO_2 pressures and lower temperatures. (From Fanale and Cannon, 1979. Copyrighted by the American Geophysical Union.)

increases adsorption by a factor of six. These and similar results were used by Fanale et al. (1982) to model exchange between the regolith and the atmosphere in response to obliquity variations. They used various models for the regolith, including mixtures of nontronite and ground basalt, and varied the total regolith thickness with latitude. They also assumed various amounts of exchangeable CO_2, but constrained this amount such that large permanent CO_2 caps do not form at present. The formation and dissipation of the permanent CO_2 caps in response to changes in insolation and atmospheric pressure were also included in the model. The calculations suggest that effects of obliquity on the atmospheric pressure of CO_2 are rather modest and are largest with the all-nontronite model, because it has a higher adsorptive capacity than basalt. At 50° obliquity the atmospheric pressure reaches 25 mbar for these models. This figure probably represents an upper limit for the obliquity effects, if the polar layered terrains do not contain large amounts of CO_2, because a thick regolith composed exclusively of highly adsorbing clays is unlikely. More probable models of the regolith, such as mixtures of ground basalt and basaltic rubble, would have significantly smaller CO_2-adsorbing capacities. Moreover, barriers to diffusive transport of CO_2 through the regolith, such as weathering horizons, duricrust or ground ice, may be significant. Some results of the modeling are shown in Figures 6-5 and 6-6.

While many details of the Fanale et al. models may be questioned, they give a good qualitative indication of what might happen during the obliquity cycle. At the lowest obliquities, most of the CO_2 is in the perennial caps and the atmospheric pressure falls to around 0.1 mbar. As the obliquity increases, more insolation falls on the poles, the perennial caps start to dissipate, and the atmospheric pressure rises accordingly. The rise in atmospheric pressure causes increased adsorption in the regolith as a consequence of the dependence of adsorption on pressure (Figure 6-4). Adsorption is greater at high latitudes because of the inverse dependence of adsorption on temperature. In effect, CO_2 is transferred from the permanent cap to the regolith. At an obliquity slightly higher than the present, the permanent caps disappear. As the obliquity continues to increase, warming of the high latitude regolith

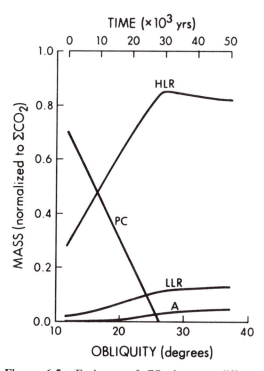

Figure 6-5. Exchange of CO_2 between different reservoirs as a function of obliquity. PC is polar cap, A is atmosphere, HLR and LLR are the high and low latitude regoliths, respectively. A 400 m thick nontronite model was assumed, which probably provides an upper limit on the amount of CO_2 that the regolith can realistically hold. (From Fanale et al., 1982. Reproduced with the permission of the Academic Press.)

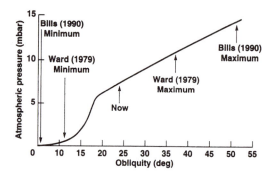

Figure 6-6. Atmospheric pressure as function of obliquity according to the model of Fanale et al. (1982) referred to in the previous figure. At low obliquities the atmosphere is buffered by permanent polar caps. Pressures at high obliquities are controlled by equilibrium with the regolith. Arrows refer to various estimates for obliquity limits. (From Kieffer and Zent, 1992. Reproduced with permission from the University of Arizona Press.)

causes desorption of CO_2 despite the still increasing atmospheric pressure, but adsorption by the low-latitude regolith continues until the maximum obliquity is reached. The scenario just described can, of course, be complicated by a variety of factors, such as changes in albedo of the cap, dust storms, and choking off the regolith with ice, but it probably is valid in general. An important point to emphasize is that unless the polar layered terrain contain large amounts of CO_2, obliquity variations are unlikely to cause pressure changes in the atmosphere of the order of 1 bar and so will cause significant greenhouse warming and allow liquid water at the surface.

The effects of the obliquity cycle on water are less clear. Ground ice may become stable everywhere at high obliquity and the stability boundary may move a few degrees poleward at low obliquities (see Chapter 2). Diffusion of water out of the low-latitude regolith will be suppressed at the higher atmospheric pressures at high obliquities and enhanced at the low atmospheric pressures at low obliquities (Toon et al., 1980; Zent et al., 1986), but these effects are minor. The main effect of increased obliquity is likely to be a major enhancement of evaporative loss from any water ice there might be at the pole, such as in a residual cap or in the polar layered terrain. Such evaporative losses could significantly increase the amount of water in the atmosphere and so change the frost-point temperature. Small changes in the frost-point temperature can significantly affect ice stablility (see Figure 2-6). Calculations indicate that at high obliquities, temperatures at the poles can get higher than 273 K and, if water is exposed at the surface, tens of centimeters of ice can sublime in a single summer (Toon et al., 1980; Jakosky and Carr, 1985). As the water-saturated air moved away from the poles the water would condense, form ice clouds, and ultimately be precipitated out. Thus at high obliquities ($>35°$) significant amounts of water could be transported equatorward to be deposited as ice in mid to low latitudes (Jakosky and Carr, 1985).

Various effects could operate to retard the process just described. If the residual cap is dusty, then a lag might accumulate on the surface and retard sublimation. The same is true, but more so, for sublimation from the polar layered deposits because they are likely to contain significant quantities of dust. Clouds may form over the summer cap and so reduce evaporation. Circulation away from the pole may be so sluggish that the humidity builds, thereby suppressing evaporation. Nevertheless, at the south pole are what appear to be ancient, partly eroded, layered deposits (Carr, 1981, p. 178). Etched into their surface are irregular hollows that resemble thermokarst pits (see Chapter 5). The erosive pattern could plausibly be interpreted as caused by loss of ice cement by sublimation during periods of high obliquity followed by deflation of the remaining silicate debris by the wind. The absence of ancient layered deposits at the north pole raises the possibility that the layered deposits there are periodically removed, possibly during periods of extreme obliquity, as a consequence of the enhanced sublimation.

Polar layered terrains

The polar layered terrains themselves occur at both poles and cover roughly the same area (1.3 $\times 10^6$ km^2). The layering is clearly visible on defrosted slopes as alternations of dark and light bands. While many of the layers can be traced for large distances, unconformities are common (Howard, 1978; Howard et al., 1982). Layering has been resolved down to 30 m, and it presumably continues at even finer scales (Figure 6-7). Suggested causes of the layering are alternating layers of dust of different density (Cutts, 1973) and variations in the dust/ice ratio (Pollack et al., 1979). From Doppler residuals Malin (1986) estimated that the layered deposits had a density of around 1000 kg m^{-3} and suggested that they must have a high ice/dust ratio. They could not be composed of dust alone, he suggested, because self-compaction would cause densities well in excess of those estimated. It is not clear, however, what errors are incorporated into the density measurement. Malin acknowledged that the errors could be significant, but calculation of formal errors was not possible. Crater counts by Plaut et al. (1988) imply that the southern layered deposits are at least several hundred million years old and have been accumulating at an average rate of no more than 8×10^{-9} km yr^{-1}. Alternatively, accumulation ceased about 10^8 years ago, leaving an upper surface that is 10^8 years old.

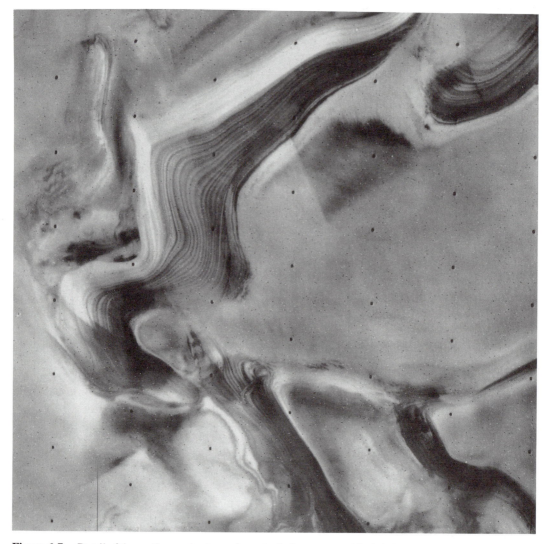

Figure 6-7.   Detail of the northern polar layered terrain during summer. The brighter areas retain CO_2 frost. The layering is seen on dark sun-facing slopes. The scene is 40 km across (Viking Orbiter frame 7B24).

Because of the rhythmic nature of the layering and location at the poles where obliquity effects are largest, the layered terrains are postulated to have accumulated in some way that has been modulated by obliquity variations (Murray et al., 1972). Cutts (1973) and Cutts and Lewis (1982) suggested that as the atmosphere thickened in response to increases in obliquity, threshold conditions for initiation of dust storms would be reached more often and dust storms would become frequent. Threshold conditions for global dust storms could be complicated, involving not only obliquity variations but eccentricity variations and the timing of perihelion. Thus periods

of deposition at high obliquities could alternate with long periods of nondeposition, and possibly erosion, at low obliquities.

Pollack et al. (1979) placed more emphasis on the role of water. They noted that present dust storms may scavenge water out of the atmosphere, and accordingly suggested that the polar deposits were mixtures of dust and ice, and that the layering was an expression of the varying ratios of dust to ice. Present deposition rates would accumulate a typical 30 m thick layer in 10^5 years, which they claimed is consistent with obliquity variations being the cause of the modulation. The crater counts of Plaut et al. (1988) indicate, however, that the layered deposits are older than formerly thought, and it appears now that the 30 m layers are unlikely to represent the 10^5 year component of the obliquity cycle, as originally suggested by Pollack et al.

More recent modeling of the exchange of water between the two poles (Jakosky et al., 1993) indicates that only very small amounts of water exchange between the two polar caps on orbital evolution time scales, a result that is consistent with the apparent old age of the deposits. They suggest that over the long term the most important factor controlling the size of any residual water caps at the poles may be the difference in elevation between the two poles. Because of the difference in elevation integrated over time, CO_2 frost tends to stay longer at the lower north pole. Because on average any southern residual water cap would be uncovered longer in summer, it loses water to the north. A balance would ultimately be achieved in which the different sizes of the residual caps on the two poles offset the differences in times that they are exposed. Thus the northern residual water cap should on average be larger than any southern residual water cap.

The layering in the polar terrains may be only partly due to the obliquity cycle. Stochastic events, such as volcanic eruptions, large impacts, and floods, must have episodically released large amounts of water into the atmosphere that would have been frozen out at the poles on time scales that are geologically short (Carr, 1990). Thus accumulation of water at the poles could have two components: the continuous slow deposition that results from desiccation of the low-latitude regolith and the rapid but episodic accumulations that result from geologic events involving water. In this case the deposits may be largely water-ice, although scavenging by dust may still be the principal depositional mechanism.

The history of the layered deposits is not just simply one of episodic accumulation. Incision of valleys into the surface of the deposits and the presence of unconformities indicates that erosion was also involved. Howard (1978) suggested that formation of the valleys is an intrinsic element in the formation of the terrain. He proposed that minor variations in the surface topography would result in preferential sublimation and deflation on sunlit slopes and condensation of volatiles on unlit slopes. Over time valleys would form, with layers being built up on the unlit slopes and being exposed and eroded away on the sunlit slopes. Thus the layering is in part a consequence of present erosional processes and not caused by addition of new material to the deposits. The crater age, according to this suggestion, is a measure of the efficacy of the recycling mechanism and not a measure of the accumulation rate.

Observational Evidence for Long-Term Climate Changes

The topic of long-term climate change is highly controversial. Some claim that Mars has never experienced major shifts in climate, others that early Mars was warm and wet but much like the present for the rest of its history, still others claim that Mars has experienced large climatic excursions throughout its history. Moreover, modeling of climate change has not led to a clear indication as to how major climatic changes might be accomplished. The observational evidence suggests that major changes in conditions at the surface have occurred, particularly at the end of heavy bombardment. Climate modeling indicates that such changes are very difficult to achieve. The two properties of the surface that suggest most compellingly that Mars has undergone major long-term climate changes are the presence of valley networks and evidence for changes in erosion rates. Several other features, such as possible strandlines, glacial features, and thick deposits of weathered debris, are also suggestive. We have already discussed outflow channels and valley networks at some length. The climatic conditions required for the formation of outflow channels are noncontroversial. The general consensus is that they formed by floods with discharges so large that they could form under present climatic conditions because freezing would be trivial compared with discharge. In fact, some outflow channels may have required a thick permafrost for their formation and conditions similar to the present in order to contain artesian water under the high pressures required to cause the huge discharges (Carr, 1979). While outflow channels may not have required different climatic conditions to form, they may have significant climatic implications. First, their formation indicates the presence of abundant water on the planet, and by implication other volatiles. Second, the floods were such enormous events that they could have temporarily disturbed global climates.

Valley networks

Valley networks are viewed as the main evidence that supports major climate changes. They have almost universally been interpreted as fluvial, that is. cut by streams of running water at the surface. If this interpretation is correct, then a convincing case can be made that climatic conditions significantly warmer than present conditions are required for their formation. The reasons that a fluvial origin may require significantly different climatic conditions are straightforward. Under present conditions with average diurnal temperatures of −60°C at the equator, precipitation of liquid water from the atmosphere is impossible and small streams would freeze, preventing the development of a network. Ice could be precipitated at high obliquities, as just discussed, but with the small climatic variations expected from obliquity changes melting of ice is possible at low latitudes, where most of the valley networks are, only on a minute scale, and under special circumstances (Clow, 1987). It almost certainly cannot provide meltwater on the scale required to produce significant streams. Thus formation of valley networks by precipitation of water or ice followed by runoff must be viewed as almost impossible under conditions resembling those that prevail at present.

It has been suggested that the valley networks could form by groundwater sapping under present conditions. The geometry of most of the networks is certainly consistent with groundwater sapping. But, as discussed in Chapter, unless the streams had very large discharges, they would freeze before they could travel the tens to hundreds of kilometers required, so that warm surface conditions are probably also required for groundwater-fed streams. Moreover, another major problem with a fluvial groundwater sapping origin for the networks under present climates is presented by the large volumes of water that are needed to cut the valleys and the need to recharge the groundwater system, as also discussed in Chapter 4. Formation of all the valleys shown in Figure 4-2 requires circulation of quantities of water equivalent to several hundred meters to kilometers, spread over the whole planet.

Such large volumes imply some recharge mechanism whereby water that passes through a valley ultimately re-enters the groundwater system so that it can re-emerge and accomplish more erosion and transport. Recharge is simple if the climate allows water to seep back into the ground, but this is not possible with a thick permafrost. Some global recharge mechanism is needed that can work despite a thick permafrost. Moreover, because most of the networks are in the Noachian, the recharge must be relatively fast, being accomplished during the late Noachian time period, possibly within 100 Myr. Recharge through basal melting of ice at the poles is possible (see Chapter 2), but the efficacy of the process is very uncertain.

As discussed in Chapters 2 and 4, hydrothermal activity could drive local circulation of groundwater, but this does not solve the recharge problem because almost all the water that is discharged onto the surface would be permanently removed from the hydrothermal circulation under present climatic conditions because the permafrost prevents its re-entry. While volcanic heat may eliminate the permafrost locally, such areas of removal are likely to be trivially small. Yet another problem with the purely sapping hypothesis is the occasional observation of relations such as initiation of streams at crater rim crests and dense drainage networks which are incompatible with purely spring-fed streams. Thus, while arguments have been made that fluvial valleys could form under present climatic conditions, it seems very unlikely. If the valleys are fluvial and close analogs to terrestrial river valleys, then surface temperatures close to or above freezing are almost certainly required.

If warm conditions are required for valley formation for the reasons just listed, the ages of the valleys present a challenge with respect to the climate history. The rate of valley formation was relatively high in the Noachian and tailed off through the Hesperian (see Chapter 4). In the Amazonian, the valleys are mostly restricted to places such as crater walls and the flanks of volcanoes, where there are steep slopes and/or high heat flows. The tail off in valley formation from the Noachian into the Hesperian could be attributed to a monotonic change in climatic conditions. Thus surface conditions may have been warm enough during the Noachian that stream flow could occur relatively often, but as conditions changed in the Hesperian the requirements for stream flow were met ever more seldom. The three to five orders of magnitude difference in drainage density between the Noachian surfaces and the Earth's surface indicate, however, that even in the Noachian conditions for surface flow were rarely met. Alternatively, the valleys were formed by some process less efficient than fluvial erosion.

The fall off in the rate of valley formation at the end of the Noachian is consistent with the drop in erosion rates discussed in the next section. If warm conditions are required to form the valleys, it could be argued that global temperatures were below but close to freezing during the Noachian, and that occasional deviations from average conditions caused temperatures to rise above freezing thereby allowing the valleys to form. Such deviations could, for example, be caused by volcanic eruptions, large floods, large impacts, or changes in obliquity and eccentricity. Decay of globally averaged temperatures through the Hesperian, it could be argued, resulted in fewer anomalously warm periods.

The Amazonian age valleys on the volcanoes present a special problem. If they formed under warm climatic conditions, why are there not more indications of such recent climate excursions elsewhere on the planet, and how did overall erosion rates remain so low throughout most of the Hesperian and the Amazonian (see later)? Possibly the combination of steep slopes, easily erodible ash at the surface, hydrothermal recycling, warm ground, and warm water allowed these valleys to form on volcanoes by fluvial action under present climatic conditions. Alternatively, large climatic excursions did occur and were short lived, as suggested by Baker et al. (1991).

The networks may not be the result of fluvial activity but the result of some other form of drainage. As discussed at length elsewhere (Carr, 1981; Baker, 1982), processes not involving liquid water are unlikely. But as discussed in Chapter 4, however, the valley networks may have formed largely by mass wasting, aided by the presence of groundwater, with the mass-wasted debris draining down valley as an ice- or water-lubricated flow. In this case, water need never flow across the surface and the quantities of water needed may be much less than with fluvial activity. The process could have been more efficient on early Mars because the higher heat flows allow liquid water at shallower depths. Such a process may be possible under present climatic conditions if heat flows were high enough, but the entire process must be considered very speculative.

In summary, if the valleys are fluvial, as is widely believed, warm surface conditions are likely to be required for their formation, both to allow the streams to flow the distances required and to re-cycle the water either through the atmosphere or the ground. The age relations suggest that the required conditions were met relatively commonly in the late heavy bombardment but much less commonly later. Young valleys on volcanoes suggest the possibility of late (Amazonian) episodic climate changes. Alternatively, the valley networks may not have such dramatic climatic implications because they are not exclusively fluvial but formed mainly by mass wasting assisted by groundwater at shallow depths.

Erosion rates

Erosion rates provide much more convincing evidence that surface conditions on early Mars were very different from subsequent conditions. In most places erosion rates have been extremely low for most of Mars history. From the survival of craters at the VL-1 landing site, Arvidson et al. (1979) estimated that erosion rates could be no more than 10^{-2} μm yr^{-1}, averaged over the mare surface at the site, which was estimated to be 3.6 Gyr old. Similar results are found in other areas. On the floors of many large craters in the uplands are preserved populations of small craters in numbers which show that the surfaces date from the base of the Hesperian (Figure 6-8). Crater production functions are commonly preserved on flat crater floors down to crater diameters at least as small as 350 m. This observation, together with measures of crater depths, indicates that the total loss of internal relief, including both erosion and filling, could not have exceeded about 60 m during the estimated 3.8 Gyr since the end of the Noachian (Carr, 1992). This gives a maximum erosion rate of 2×10^{-2} μm yr^{-1}, a result consistent with that of Arvidson et al. The actual erosion rate could have been significantly less because deposition of dust must have contributed to crater obliteration.

The results just described apply to bedrock units only. Surficial deposits such as the wind-de-

Figure 6-8. The floor of a large upland crater at 0°, 259°W. The number of craters on the floor indicates that the floor dates from the Noachian-Hesperian boundary. Despite a 3.5–3.8 Gyr age for the surface (see Table 1-1), these small craters have undergone very little degradation, as is typical for low latitude post-Noachian surfaces. Contrast this with the amount of degradation sustained by large Noachin craters in Figures 4-1 and 4-4. The scene is 17 km across (Viking Orbiter frame 130S09).

posited drifts at the Viking landing sites are clearly eroded and deposited at much faster rates. In many places, older surfaces are unconformably overlain by what are seemingly easily erodible deposits. The most extensive such deposit is the Medusa Fossae Formation found just south of the equator between 140°W and 220°W, but numerous patches are found elsewhere. The contrast in erosion between these materials and the underlying surface suggests a great contrast in erodibility. Although other origins have been suggested, these materials are plausibly interpreted as composed of eroded and weathered debris that formed during the periods of high erosion during the Noachian and that have since that time been moved around the surface of the planet by the wind.

The contrast between the Noachian and younger surfaces is dramatic. The cratered upland surfaces (see Figures 4-1, 4-4, and 4-21) show craters in sizes up to at least 70 km in diameter in almost all stages of degradation, and the number of smaller craters (<20 km diameter) is disproportionately low compared with the larger craters. The significance of the disproportionately low number of small craters was recognized early. Hartmann (1973) suggested that as a result of high obliteration rates during heavy bombardment, few of the smaller craters survived from that era. When heavy bombardment was over and obliteration rates declined, a younger population of small craters was superimposed, so that the present size frequency distribution is a result of the change in crater obliteration rates near the end of heavy bombardment.

The degradation of the larger craters appears not to be due to isostatic readjustment because whereas the high frequency components of the relief, such as crater rims and ejecta textures, are removed, there is little evidence of removal of the main low-frequency components by bowing up of the crater floors. Nor does the degradation appear to be simply a result of modification of the surface by the cratering process itself, for the craters have undergone much more degradation than in the lunar and mercurian highlands despite being protected from impact erosion by the smaller bodies by the atmosphere. Erosion rates are difficult to estimate, but, assuming impact rates similar to the late heavy bombardment on the Moon given in Wihelms (1986), I estimated that the erosion rates were on the order of 10 μm yr^{-1} (Carr, 1992), three orders of magnitude higher than rates estimated by Arvidson et al. (1979) for the post-Noachian. For reference, denudation rates on arid regions of the Earth are estimated to range from 10 to 1000 μm yr^{-1} (Saunders and Young, 1983). There are other indications of a very rapid fall-off in erosion rates at the end of the Noachian. Baker and Partridge (1986) point out that the presence of Noachian valley networks in very different states of preservation implies that erosion rates were falling rapidly at the end of the Noachian. Those that formed right at the end of the Noachian have undergone almost no erosion since they formed around 3.8 Gyr ago, while those that formed somewhat earlier, say, 4 Gyr ago, are highly eroded. Schultz and Britt (1986), on the basis of drainage densities on Noachian units with different crater ages, also demonstrated that degradation rates dropped rapidly at the end of the Noachian, between the formation of the Isidis basin and the Argyre basin.

Craddock and Maxwell (1993) made a detailed assessment of how degradation rates changed with time, location, and elevation at the end of the Noachian. According to the time scale of Scott and Tanaka (1986), a surface dating from the end of the Noachian (or the base of the Hesperian) should have a crater density of 200 craters larger than 5 km per 10^6 km^2. If erosion rates have been negligible since the Noachian but were higher during the Noachian, then Noachian surfaces should have the basal Hesperian numbers of fresh craters (200 > 5km per 10^6 km^2) but significantly larger total numbers of craters. This is in general what Craddock and Maxwell found, again confirming the rapid dropoff in erosion rates at the end of the Noachian. There were, however, some slight modifications to this simple picture. Not surprisingly, they found that the larger the crater, the further back into the Noachian its fresh appearance could be preserved. But more difficult to understand is that the lower the altitude, the longer that high erosion rates appear to have been sustained, although even at the lowest elevations the high erosion rates terminated in the lower Hesperian. Craddock and Maxwell attributed the crater degradation mainly to fluvial erosion following precipitation. Their estimates of 0.1–5 μm yr^{-1} of the erosion rates during the late Noachian are somewhat less than my estimate of 10 μm yr^{-1}.

The substantial fall in erosion rates at the end of heavy bombardment is thus well substantiated and cannot be doubted. Its cause is not so clear, although it is difficult to see how an atmospheric cause can be avoided.

Other possible evidence for climate change

While the valley networks and erosion rates present the strongest evidence for climate change, other observations have been cited in support of climate change. The first is the presence on the surface of thick accumulations of what are probably the products of erosion and weathering. The Medusae Fossae formation, the most prominent example, has already been mentioned, but smaller patches of similar deposits occur elsewhere. Where present, the surface has an etched appearance, pedestal craters are common, and mounds of seemingly layered deposits occur within large craters (Schultz and Lutz, 1988). The two largest areas outside the Medusae Fossae Formation outcrops and surroundings are a broad region northwest of Isidis Planitia and on the equator near the zero longitude. They are mapped as etch deposits on the 1:15 M geologic maps (Scott and Tanaka, 1986; Greeley and Guest, 1987). While alternatives origins have been suggested for these deposits, they can plausibly be interpreted as accumulations of eroded and weathered material. Drifts at the Viking landing sites, possible debris blankets on the northern plains (Soderblom et al., 1973), and extensive dune fields, particularly around the north pole and locally at high southern latitudes (Greeley et al., 1992) are further evidence for abundant loose material at the surface. Given the extremely low erosion rates since the end of the Noachian, these materials are unlikely to have been produced during this time. They are more likely the product of weathering and erosion during some early period in the planets history when climatic conditions were more conducive to such processes. There being no oceans to capture such materials, they have since merely been redistributed around the planet in response to changes in wind regimes (Figure 6-9).

Two additional more speculative suggestions made in support of climate change, the former presence of glaciers and oceans, have already been discussed in previous chapters. The confluence of a number of landforms resembling terrestrial glacial features, such as eskers, moraines, kettles, glacial scour, cirques, and arretes, particularly in and around Hellas and Argyre, suggested to Kargel and Strom (1992) that Mars has experienced alpine-type glaciation late in martian history, during which as much as $1-7 \times 10^7$ km^3 of ice might have accumulated at high southern latitudes. If so, then a warm, thick atmosphere is needed to provide the precipitation. Kargel and Strom ascribe the glacial episode to one or several climatic excursions that could have lasted 10^5 to 10^7 years. The possibility that the northern lowlands formerly contained ocean-sized bodies of water was discussed in Chapter 3. If oceans

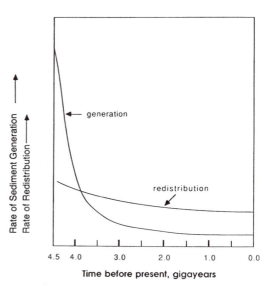

Figure 6-9. Diagram of the generation and redistribution of dust on Mars. Much of the fine grained material presently at the surface may have formed by weathering and erosion very early in the planet's history, then redistributed with time, with only modest additions during the last 3 Gyr (From Gooding et al., 1992. Reproduced with permission from the University of Arizona Press.)

were present and the strandlines do indeed mark former shorelines, then climate change is almost certainly implied, as it is difficult to see how such large bodies of water could dissipate under present conditions and leave behind strandlines, outlining the now empty ocean basins. Clearly, these last two suggestions are very speculative.

Geologic and Geochemical Estimates of the Surface Inventory of CO$_2$

Almost all climate and geologic models for the evolution of Mars postulate fixation of a thick CO$_2$ atmosphere as carbonates. Yet searches in the visible and near-infrared for spectral indications of extensive carbonate deposits have been largely unsuccessful (see summary in Blaney and McCord, 1989). For example, high-wavelength resolution spectra taken of four broad areas (Syrtis, highlands north of Hellas, Hellas, and Arabia) obtained by Blaney and McCord during the excellent 1986 opposition showed that no more than 3 weight percent calcium carbonate or 5 wt % magnesium carbonate could be present in the upper few millimeters in these regions. While dust could easily mask surface deposits, the results suggest that the surface dust itself contains no more than a few percent anhydrous carbonates. More recently, reinterpretation of *Mariner 6/7* spectra and new airborne infrared measurements may have resulted in the detection of larger amounts of carbonates. Calvin et al. (1994) suggested that a feature at 4.5 μm, interpreted as caused by surface or atmospheric dust adsorption, could result from the presence of hydrous magnesium carbonates. Such an interpretation is consistent with other features near 2.3 μm identified in Earth-based spectra. They suggest 10–20% by weight is consistent with both the spectral measurements and the Viking soil analyses.

Pollack et al. (1990), observing from the Kuiper Airborne Observatory, obtained spectra in the 5.4–10.5 μm region of several large spots on the martian disc. They identified several adsorption and emission features caused by dust in the atmosphere and on the surface. The features were tentatively identified as due to silicates, molecular water, sulfates or bisulfates, and carbonates or bicarbonates. From the strength of the sulfate and carbonate lines, they estimated that sulfate and carbonate abundances in the dust were approximately 10–15% and 1–3%, respectively. They noted that if carbonates were in the ground in these proportions and the upper 10 km of the megaregolith was well mixed by volcanic and impact processes (Pollack et al., 1987; Carr, 1989), then the regolith could contain about 5 bars of CO_2.

Table 6-2. Products of Aqueous Alteration in SNC Meteorites.

(From Gooding, 1992)

	Shergottite	Nakhla	Chassigny
$CaCO_3$	x	x	x
Mg-bearing $CaCO_3$	x		
$MgCO_3$			x
$(Fe,Mn)CO_3$		x	
$CaSO_4 \cdot nH_2O$	x	x	x
$(Mg)_x(PO_4)_y \cdot nH_2O$	x		
$(Mg)_x(SO_4)_y \cdot nH_2O$		x	
$(Na,K)Cl$		x	
"illite" $(K,Na,Ca_{0.5},H_3O)(Al,Mg,Fe)_2$ $(Si,Al)_4O_{10}[(OH)_2,H_2O]$	x		
S, Cl-bearing "micabole"	x		
Smectite $(Na,Ca_{0.5})_{0.3}(Al,Fe,Mg)_{2-3}$ $(Si,Al)_4O_{10}(OH)_2 \cdot nH_2O$		x	
$Fe_2O_3 \cdot nH_2O$		x	

Water soluble salts, specifically carbonates, sulfates, and halites, have been detected in small amounts (<0.1%) as secondary alteration products in SNC meteorites (Table 6-2) (Gooding et al., 1988, 1991; Gooding, 1992). The mineralogy suggests chemical alteration at temperatures in the 270-373°K range by oxidizing solutions with pH >8. A newly discovered SNC meteorite (ALHA84001) has larger amounts of carbonate (Mittelfehldt, 1994), which has a composition suggestive of alteration at moderately high temperatures (700°C) and hence of hydrothermal activity. Thus, there is excellent mineralogic evidence for circulation of CO_2-rich fluids in the crust, as expected if there are carbonates in the megaregolith.

Griffith et al. (1995) point out that hydrothermal systems can be quite efficient in fixing CO_2. In terrestrial hydrothermal systems, carbonates are dispersed in vugs, veins, and mineral alteration products so are not accessible for detection at the surface. In Iceland, carbonates formed by the alteration of basaltic rocks by volcanically warmed meteoric waters have been detected from the surface down to depths of 2 km in the deepest wells. A conservative estimate of the total CO_2 sequestered in Iceland along is 2.4 mbar compared with only 0.4 mbar in the Earth's entire atmosphere. All the 2.4 mbar has been sequestered in the geolocially short time of only 16 Myr. Given the sustained volcanism on Mars, the likelihood that groundwater is charged with CO_2 by hydrothermal activity could have been very significant on Mars also.

Several geologic observations also suggest the presence of carbonates dispersed in the near surface materials. Evidence of subsurface erosion and/or solution has been referred to in several places in previous chapters, particularly in reference to formation of the outflow channels (see Figures 3-4 and 3-7) and fretted terrain (Figure 4-14). The formation of such features is readily explained if the megaregolith contains a few to several percent of soluble salts. Groundwater movement would then lead to solution and disaggregation of the megaregolith materials, thereby facilitating the formation of the underground conduits implied by the relations in the above-mentioned figures. If early Mars did have a thick CO_2 atmosphere, then mixing of carbonates (or bicarbonates) into the megaregolith by volcanic burial and impact gardening, as discussed later is to be expected.

Spencer and Fanale (1990) additionally pointed out that many of the characteristics of the canyons, particularly the large, closed depressions (Figure 1-11), can be explained on the basis of a solution of massive carbonate deposits. They postulated that during heavy bombardment accumulation of carbonates was preferentially concentrated at the equator, where warmer conditions might have favored the presence of liquid water. They suggested that thick carbonate deposits could have formed in the regions of the canyons during heavy bombardment and that these were later partly dissolved by groundwater to form the karstlike features observed. While these geologic observations and suggestions do not lead to any quantitative estimates, they do indicate that carbonates could be present, either as discrete deposits or dispersed through the megaregolith.

If the H_2O/CO_2 ratio on Mars were comparable to the ratios for Earth, for which estimates range from 1.8 to 8 (McElroy et al., 1977; Pollack and Black, 1979; Walker, 1977), then geologic estimates of 400 m of water at the near surface, based on erosion of outflow channels (Carr, 1986), would imply 2–10 bars of CO_2 are present. Although such an assumption has been made in the past to estimate the amount of CO_2, the resulting estimates are questionable, because the assumption of similar ratios implies that all the various exchange and loss mechanisms that have operated on the planet from accretion through to the present have similarly fractionated both species with respect to each other on both planets, which is quite improbable.

Mechanisms for Climate Change

Most modeling of past climates on Mars has been concerned with what kind of an atmosphere had to be present toward the end of heavy bombardment in order to elevate surface temperatures and so allow liquid water to flow across the surface and cut the valley networks. Attention has focused on CO_2–H_2O atmospheres because of the susceptibility of other plausible greenhouse gases, such as NH_3, CH_4, and SO_2, to UV-induced photodissociation or chemical reaction. Not too much attention has been paid to how a thick CO_2–H_2O atmosphere might initially form, although, because of its vulnerability to collapse through the formation of carbonates, several investigators have examined how such an atmosphere might be sustained from the time of formation of the planet around 4.5 Gyr ago to the end of the Noachian, around 3.8 Gyr ago.

Greenhouse warming

Optically active gases within the atmosphere produce surface warming by allowing most of the UV and visible solar radiation to pass through, but absorbing some significant fraction of the long-wave radiation emitted by the surface and lower atmosphere. At Mars' current temperatures, the part of the infrared spectrum of concern is from 7 to 100 μm (Pollack, 1979; Fanale et al., 1992). The main CO_2 absorbtion band in this range is the 15 μm band. Water is a strong absorber in the 8–12 μm range, but at current low temperatures the water content of the atmosphere is so low that its effect is small. As surface temperatures rise under a thicker CO_2 atmosphere, more water can enter the atmosphere and it becomes more important as a greenhouse gas. Ammonia is a potent greenhouse gas because of its 10 and 16 μm bands and rotational transitions longward of 20 μm. Methane is less effective but can produce significant warming at total atmospheric pressures upward of 1 bar. Hydrogen has rotational lines at 17 and 28 μm, and its absorption also becomes significant at high pressures. Sulfur dioxide, with vibrational transitions near 7, 9, and 19 μm and rotational transitions at longer wavelengths, is another potentially strong greenhouse contributor. All these greenhouse gases become more effective as increased abundances cause weak lines to become significant and as increasing the total pressure causes broadening of all the lines.

Several different attempts have been made to model greenhouse warming of Mars under a CO_2-H_2O atmosphere (Pollack, 1979; Cess et al., 1980; Hoffert et al., 1981; Postawko and Kuhn, 1986; Pollack et al., 1987). These models differ mainly in how they calculate temperature profiles, the humidities that they adopt, the complexity with which they treat the infrared absorption of the atmosphere, and the degree to which they incorporate latitudinal differences. The models are in general mutually consistent. Pollack et al. (1987) used a one-dimensional model, in which solar and thermal heating is radiatively balanced throughout the atmospheric profile, except that to

account for convection in the troposphere, the lapse rate was set to the moist adiabat whenever the radiatively derived rate exceeded the adiabatic rate. Humidity was assumed to vary linearly with pressure, starting at 0.77 at the surface. Improved models of the adsorption of CO_2 and H_2O over previous models were incorporated. The results are summarized in Figure 6-10. With the present solar constant and a surface albedo of 0.215, a surface pressure of 2.2 bars is needed to raise the surface temperature to the melting point of ice. However, the solar luminosity is predicted to have been only 0.7 times the present 4.5 Gyr ago and 0.75 times the present 3.8 Gyr ago (Newman and Rood, 1987; Gough, 1981). At 0.75 times the present luminosity, a 5 bar atmosphere is needed for the surface to reach 273 K. The requirements are somewhat less for a surface albedo of 0.1. The Pollack et al. results are globally averaged. Latitudinally resolved calculations (Postawko and Kuhn, 1986; Hoffert et al., 1981) show similar results. It is worth noting in passing that valley networks are found at high latitudes, so that warming at all latitudes is required for a fluvial explanation of the channels.

Subsequently, Kasting (1991) suggested that these results might be in error because they did

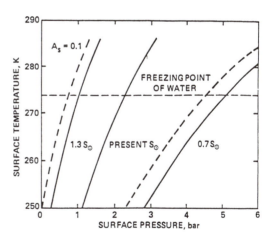

Figure 6-10. Surface temperature as a function of surface pressure with a CO_2–H_2O atmosphere for several values of the surface albedo (A_s) and the incident solar flux, S. The solid lines are for the present globally averaged albedo of 0.215. Thus for the present solar flux, 2-3 bars of CO_2 are needed to bring surface temperatures up to freezing, according to this model. (From Pollack et al., 1987. Reproduced with permission from Academic Press.)

not incorporate the effects of CO_2 condensation. This significantly reduces any greenhouse warming and makes it difficult or impossible to raise the global surface temperature above 273 K with a CO_2–H_2O atmosphere alone. The radiative-convective model just described results in an atmosphere greatly oversaturated with CO_2 between altitudes of 20 and 40 km, which is physically impossible. Correcting the temperature profile for condensation in this region causes the surface temperature to drop about 9 K for present solar luminosities. For luminosities below 0.9, surface temperatures and pressures reach a limiting value because of "runaway condensation." Reducing the solar constant causes lower temperatures throughout the atmosphere and causes more CO_2 to condense. Increased condensation lowers the tropospheric lapse rate, thereby lowering the surface temperature and the altitude at which condensation occurs. For a solar constant of 0.7, that expected for early Mars, the surface temperature reaches a maximum of 214 K at a surface pressure of 1.5 bars (Figure 6-11). A thicker atmosphere does not result in an increase in temperature, because any increase in infrared adsorption of the atmosphere is offset by increases in the planetary albedo that result from the thicker atmosphere. At low surface pressures both the greenhouse effect and planetary albedo are small; at high surface pressures both are large. The maximum warming occurs at some intermediate pressure, dependent on the solar constant. According to these calculations, a CO_2–H_2O greenhouse alone could not have resulted in surface temperatures such that liquid water could flow if the solar constant was 0.7 times the present, as is expected from stellar evolution models. The results may be even more discouraging if latitudinal mixing is included.

These calculations create a major dilemma, as discussed by Kasting (1991). If the channels are fluvial then temperatures above 273 K appear required both to allow surface flow and to recharge the groundwater system, and the unequivocal high erosion rates in the late Noachian also need to be explained. Kasting suggested various ways of improving the greenhouse models, such as better modeling of both CO_2 and H_2O clouds, CO_2 absorption at high pressure, and the non-ideal equation of state of CO_2, but it is not clear that these improvements in the modeling will alleviate the problem caused by CO_2 condensation.

Trace gases, such as NH_3, CH_4, and SO_2, might enhance the greenhouse, but their stability would have to be explained in light of their probable rapid photodissociation and chemical reactions (Kuhn and Atreya, 1979; Yung and Pinto, 1978). NH_3 is photodissociated by the Sun's UV radiation, and the hydrogen produced is lost to space. On early Mars the solar UV output is expected to be higher (Zahnle and Walker, 1982), and loss rates are likely to be such that 1 mbar of NH_3 would be lost in less than 100 years and there is no plausible replenishment mechanism. Similarly, photochemical reaction could convert 1 bar of methane to higher order hydrocarbons in less than 100 years (Yung and Pinto, 1978). On the other hand, the stability of both NH_3 and CH_4 could be enhanced by the formation of a hydrocarbon smog from methane photolysis (Squyres and Kasting, 1994), and modest amounts of NH_3 could be maintained by photochemical reduction of nitrate or nitrite produced by lightning (Summers and Chang, 1993). Sulfur gases injected into the atmosphere are likely to be rapidly converted to sulfuric acid particles if water is present, so these gases also would have to be rapidly replenished. Large amounts of suspended, impact-

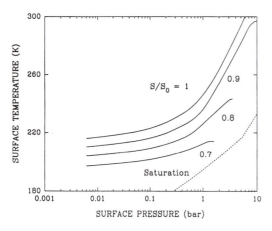

Figure 6-11. Surface temperature versus surface pressure with a CO_2–H_2O atmosphere for four different values of the solar luminosity. The dashed curve shows the condensation vapor pressure of CO_2. For luminosities below 0.9 time the present Sun, increases in CO_2 pressure do not result in higher temperatures because of condensation to form clouds and the consequent increase in the planetary albedo. (From Kasting, 1991. Reproduced with permission from Academic Press.)

generated dust could affect surface temperatures, but they are more likely to have a cooling effect than a warming effect. Erosion of the channels by liquid CO_2 is marginally possible. The CO_2 triple point is at 216.56 K and 5.11 bars, so large amounts of CO_2 would be needed. Erosion by liquid hydrocarbons under a methane atmosphere is also possible, but this is difficult to believe in view of the susceptibility of methane to UV-induced photochemical reactions as just discussed.

Another possibility, discussed by Kasting (1991), is that the solar constant early in Mars' history was larger than stellar models predict (Gough, 19981). A delicate balance must be achieved because Mars must be warmed to above 273 K without raising the amount of sunlight needed such that the Earth's oceans evaporate. The solar luminosity models would be in error if the Sun continued to lose mass after it reached the main sequence. Boothroyd et al. (1991) have invoked a 10% mass loss to explain the ^{7}Li depletion. Luminosity scales as $M^{4.75}$ and planetary orbits scale as $1/M$, so that a 10% mass loss would increase luminosity by 1.9, which, given the 30% drop in luminosity in the last 4 Gyr, results in a luminosity 1.33 times the present value. This is enough to warm Mars but may create a water loss problem for the Earth (Kasting, 1987). Kasting (1991) suggested that a solar mass a few percent larger than the present mass would be optimum, permitting early Mars to be warm and at the same time solving the "early, faint, young Sun paradox" for the Earth. This paradox results because if the composition of the Earth's atmosphere was the same as at present, and the luminosity of the Sun was as low as is postulated, the Earth's oceans would have been frozen prior to around 2.5 Gyr ago, which is contrary to the geologic record (Sagan and Mullen, 1972). The Earth's atmosphere must have contained higher proportions of some greenhouse gas(es), and CO_2 is the most likely candidate (Owen et al., 1979). Assessing the plausibility of a somewhat larger, early Sun must await a better understanding of the evolution of main sequence stars.

In summary, current climate models cannot be satisfactorily reconciled with the concept of an early, warm and wet Mars. Either early Mars was not warm and wet, or there is some fundamental, unrecognized fault with the climate models.

Maintenance of an early thick atmosphere

Impact erosion would work to eliminate an early thick CO_2–H_2O atmosphere if it formed. Large impacts can create a hot vapor plume that expands outward with sufficient energy to blow off part or all of the overlying atmosphere. The first condition to be met for significant atmospheric erosion to occur is that the impact velocity must be large enough to create a plume that will expand faster than the planet's escape velocity. Melosh and Vickery (1989) estimated that this threshold velocity is 14.3 km sec^{-1} for silicate impactors and 11.1 km sec^{-1} for icy impactors. The second condition to be met is that the mass of the vapor from the projectile and target must exceed the air mass above the plane tangent to the point of impact. Melosh and Vickery estimated that for this conditions to be met at the present time, an object about 3 km in diameter is needed. Melosh and Vickery then used the lunar record to estimate the cratering rate on Mars during heavy bombardment and estimated that about 100 times the present mass of the atmosphere had been removed (Figure 6-12). In their model, Mars had a 1 bar atmosphere at the end of accretion, and this rapidly declined soon after the end

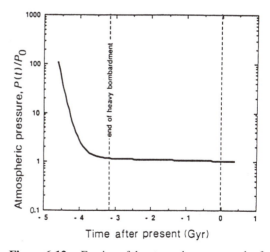

Figure 6-12. Erosion of the atmosphere as a result of impacts. According to this model, any atmosphere present at the end of accretion is rapidly eroded away during the first half of the heavy bombardment period as a consequence of blow-off by large impacts. (From Melosh and Vickery, 1989. Reproduced by permission from Nature. Copyright 1989, Macmillan Magazines Ltd.)

of accretion so that around 4 Gyr ago the atmospheric pressure was only a few tens of millibars.

Although Melosh and Vickery claimed that this result is consistent with observational evidence for fluvial features and high erosion rates at the end of heavy bombardment, the claim is questionable. The Noachian record preserved at the surface must be biased toward the end of heavy bombardment, yet we see abundant evidence of network formation and high erosion rates. The rate of these processes dropped at the end of heavy bombardment, not at the end of accretion, as is implied by the Melosh and Vickery model. Even in their model, Mars had to be able to accumulate a thick atmosphere at the end of accretion, when impacts rates were presumably much higher than they were toward the end of heavy bombardment. Melosh and Vickery assumed present impact velocities, but impact velocities could have been lower at the end of accretion and early during heavy bombardment, thereby undermining the efficacy of the impact erosion process.

If early Mars was warm and wet, then chemical weathering would have tended to consume the atmospheric CO_2 and convert it to carbonates. On Earth CO_2 dissolves in rain and surface waters to form carbonic acid, which reacts with surface rocks. Cations such as Mg^{2+} and Ca^{2+}, and HCO_3^- anions are transported to the ocean to be deposited as carbonates, generally through mediation by living organisms. The scavenging process is efficient. All the CO_2 in the Earth's atmosphere-ocean system is removed on a time scale of 4×10^5 years (Holland, 1978), but it is replaced by volcanic emanations so that a rough steady state is achieved. Pollack et al. (1987) attempted to model what might happen on Mars if it had warm conditions under a CO_2 atmosphere. They attempted to scale terrestrial weathering rates to Mars by appropriately accounting for different conditions on the two planets. They assessed the effects of five factors: CO_2 pressure, the variation in runoff with temperature, the variation in reaction rates with temperature, the cation content of the rocks at the surface, and the fraction of the surface covered by open bodies of water, which is the ultimate source of the precipitation providing the runoff. The main uncertainty in the calculations is estimating the last factor, the amount of open water available. Assuming a fraction of open water was 0.05 times that on the

Earth and a surface temperature of 273°K, they estimated that a CO_2 atmosphere of a bar would be eliminated in roughly 10^7 years. However, the value of 0.05 for the scaling factor for open water may be too high in view of the low drainage densities observed in the martian uplands. The atmosphere may be thus more resilient to collapse than these numbers indicate.

Schaefer (1990) did similar, but more detailed, calculations of how varying the CO_2 pressure in the presence of water affected equilibria in the calcium carbonate-anorthite-kaolinite system in order to determine how weathering might remove CO_2 from the atmosphere in the presence of large bodies of water. On the basis of the modeling, she suggested that thick carbonate deposits should have formed in the northern plains if large bodies of water had been present there in the presence of a thick CO_2 atmosphere.

Recharge of the atmosphere with CO_2

Loss of CO_2 from the atmosphere would be counteracted by resupply as a result of volcanism. Volcanism could both outgas juvenile CO_2, thereby adding to the near-surface inventory, and recycle CO_2 fixed as carbonates by burial to depths where dissociation takes place. Plausibly assuming that early Mars would have had high rates of volcanism because of the high heat flows, Pollack et al. (1987) estimated the rate at which carbonates might be recycled on early Mars as a result of volcanic burial. Large errors are not likely in the estimated heat production (Schubert and Spohn, 1990). The main uncertainties are the thermal conductivity of the crust, which affects the depth to which carbonates must be buried in order for them to dissociate and the fraction of the heat flow that is advected to the surface as lavas, as opposed to that conducted through the lithosphere. The best estimate of Pollack et al. is that the recycle times of CO_2 are on the order of 10^8 years. In addition to decomposition by burial, volcanic activity could release CO_2 into the atmosphere more directly. Schaefer (1993) noted that eruption of lavas onto the surface would decompose any carbonates present and release CO_2 into the atmosphere. Large eruptions of fluid lavas onto the carbonate deposits that she postulates for the northern plains would efficiently return CO_2 into the atmosphere.

How ?

Despite the earlier discussion on impact erosion, impacts may have aided in the recycling of CO_2 (Carr, 1989). The coincidence in the timing of the drop in impact rates with the drop in rates of erosion and valley formation suggests a connection and that somehow the high impact rates were helping maintain the required conditions. Three impact-related processes could have contributed CO_2 to the atmosphere: impact gardening, that is, stirring the near-surface materials by repeated impacts, shock-induced decomposition of carbonates, and accretion of meteoritic carbon. Crude calculations, using impact rates comparable to lunar rates estimated for the end of heavy bombardment, suggest that impacts alone could not recharge the atmosphere with CO_2 at the rate of 1 bar every 10^7 years estimated by Pollack at el. (1987). However, if they overestimated the fraction of open water on early Mars, as appears likely, and the fixation rate was closer to 1 bar every 10^8 years, then impact-caused recycling could be significant. For it to work, however, impact velocities during heavy bombardment would have to be lower on average than they are today to offset the effects of impact erosion. This is not unreasonable. The objects that caused the heavy bombardment were a distinctively different population from those that caused subsequent impacts (Strom et al., 1992). They may represent the remnants of the material in the inner solar system that accumulated to form the planets and had more circular heliocentric orbits and hence lower impact velocities than typical present day impactors.

Episodic ocean hypothesis

Evidence for Amazonian fluvial activity and for similar-aged, possible glacial features at high southern latitudes suggested to Baker et al. (1991) that the conventional view of a warm, wet early Mars changing early to the frigid Mars of today was untenable and that a new paradigm for the geologic history of Mars was needed. These observations, coupled with interpretations that linear features outlining large areas of the northern plains are fossil shorelines (Parker et al., 1989), led Baker et al. (1991) to propose a scheme in which a northern ocean, which they called the Oceanus Borealis, repeatedly formed and dissipated in Hesperian and Amazonian times, temporarily producing warm and wet climates. The oceans formed as a consequence of the large floods. They acknowledge that for most of Mars' history, after the end of the Noachian Mars was much like it is today, with a thick, possibly ice-rich, permafrost overlying a permeable, and probably water-rich, upper crust. They suggested that episodically massive volcanic activity in the Tharsis region caused widespread melting of ground ice, thereby releasing large amounts of water into the groundwater system and triggering groundwater eruptions and floods around the Chryse basin. Discharges estimated for these floods are in the range of 10^7 to 10^9 m^3 sec^{-1} (see Chapter 3). With these discharges, to form oceans of the volumes they propose (Table 6-3, many floods would need to occur almost simultaneously. This is not difficult to imagine, in that the individual floods are so large that they would have caused massive disruption and decompression of the aquifer system, so that one flood might readily trigger another. The times estimated to create the oceans range from 10 days to 2 years depending on the size of the ocean and the discharge rates.

Baker et al. suggested that such massive releases of water and the formation of the resulting oceans would have had major climatic implications. If early Mars had a thick CO_2 atmosphere, then the ground may contain large amounts of carbonates or CO_2-containing compounds such as clathrates. Groundwater is thus expected to have been charged with CO_2, and this would have been released during the floods. Additional sources of CO_2 are the regolith and volcanic activity. The massive release of CO_2 and H_2O into the atmosphere, they hypothesize, caused global warming, promoting the release of more of these gases from clathrates in the regolith, the polar caps, and polar layered terrains. In order for temperatures to rise so that liquid water could be stable at the surface, approximately 3 bar of CO_2 would need to be released (Pollack, 1979). Because we are discussing later epochs, the faint Sun problem discussed in the previous section is not an issue. Surface temperatures in excess of 273 K are needed in their hypothesis to provide precipitation to sustain the southern glaciers and to erode the young channels. They estimate, for example, by analogy with Hawaii, that to cut the channels on Alba Patera requires the equivalent

Table 6-3. Hypsometry of Mars northern plains (Oceanus Borealis)

(From Baker et al., 1991)

Contour[*] (km)	Volume (10^7 km^3)	Area (10^7 km^2)	Average depth (km)	Equivalent[+] water layer (m)
0	6.5	3.8	1.7	450
-1	3.1	2.8	1.1	210
-2	1.0	1.4	0.7	70

[*] Water below the stated contour.   [+] Covering the entire martian surface.

of a column of water 250 km high to pass through the channels in 10^5 years. The oceans could not accumulate slowly, because under present climatic conditions each small increment added would freeze. The climate forcing must be primarily by CO_2, and very large amounts are needed.

After the oceans formed, CO_2 would start again to be fixed as carbonates and the ocean floor would warm, thereby allowing the water to infiltrate back into the underlying megaregolith. The oceans would therefore disappear to recharge the groundwater system, the thick CO_2 atmosphere would dissipate, and dry, frigid surface conditions would return until volcanism triggers the next sequence of massive floods. But in the short interim, the oceans would have provided sufficient precipitation to cut the young channels and cause the young glaciations. Carbon dioxide presents a particularly difficult problem for the episodic ocean hypothesis. If large bodies of water formed in the presence of a thick CO_2 atmosphere, formation of carbonates is likely (Schaefer, 1990). In the absence of high rates of volcanism and impacts, it is difficult to see how the CO_2 locked up in these deposits could be recycled in order for the CO_2 to participate in the next oceanic episode.

This scenario was invoked to explain, in one self-consistent hypothesis, several otherwise difficult to explain observations. If fluvial, the young valley networks, like those on Alba Patera, imply warm climatic conditions late in martian history. The glacial-like landforms, such as those in southern Hellas and Argyre, if they are

truly glacial, require substantial amounts of precipitation. The linear features in the northern plains, if shorelines, imply the presence in post-Noachian times of large bodies of water. Thick sediments in the canyons appear to require long-lived lakes in post-Noachian times. All these observations must be explained in a manner consistent with the extremely small amounts of erosion that have accomplished over the last 3.8 Gyr. Baker et al. envisage a history of Mars that consists of long periods of low geomorphic activity, with climatic conditions much as they are today, separated by periods of intense geomorphic activity during several short, warm, wet maritime episodes.

The plausibility of the model depends to a large extent on the plausibility of the geomorphic interpretations that stimulated formulation of the model. The young valley networks, the postulated glacial features, the young lacustrine sediments, and the postulated shorelines all have alternative interpretations, in addition to those accepted by Baker et al. (1991). As discussed in previous chapters, some of these interpretations do not demand climatic conditions greatly different from the present. The advantage of the Baker et al. hypothesis is that all these anomalies are explained within a single self-consistent model. The disadvantage of the model is that many of the proposed processes are very poorly understood.

Summary

The obliquity of Mars varies chaotically between roughly 0° and 60°. At large obliquities more insolation falls on the poles than on the equator, and volatiles such as H_2O and CO_2 will tend to be driven off. CO_2 trapped in the permanent southern cap and the high-latitude regolith could raise surface pressure 10–20 mbar at most. If the polar layered deposits contain large amounts of clathrate or solid CO_2, for which there is no evidence, surface pressures could increase to several tenths of a bar, still insufficient to raise surface temperatures to above freezing by greenhouse warming. Valley networks and erosion rates provide the most compelling evidence for long term climate change. To form valley networks by fluvial processes almost certainly requires surface temperatures above freezing. Erosion rates dropped from values close to terrestrial rates during heavy bombardment to values at least three orders of magnitude lower shortly after the end of heavy bombardment. These low rates have been sustained for much of Mars' subsequent history. Models of surface evolution must account not only for the early high rates of erosion and valley network formation, but also the less common younger networks. Warm surface temperatures are difficult to achieve with CO_2–H_2O atmospheres, particularly early in the planet's history when the luminosity of the Sun was low. Even with present luminosities, 2–3 bars of CO_2 are needed. A thick CO_2–H_2O atmosphere is unstable and difficult to maintain because of carbonate formation, but high rates of volcanism and impact early in the planet's history may have helped support a thick atmosphere. Small fractions of carbonates have been detected in the loose material at the surface and in the atmosphere, but large carbonate deposits have not been detected. Raising surface temperatures by means of other greenhouse gases, such as NH_3, CH_4, and SO_2, is difficult because of their short lifetimes within the atmosphere. Current climate models cannot explain how early Mars could have been warm and wet. Either the models have some fundamental flaw or early Mars was not warm and wet. A controversial proposal has been forwarded to explain relatively young, supposed glacial features, young valley networks, and evidence of former bodies of water in the northern plains. The suggestion is that large floods episodically created ocean-sized bodies of water. These temporarily changed the global climate and then rapidly dissipated. However, the precise mechanisms whereby the climate changed, the oceans dissipated, and the CO_2 fixed in carbonates was released so that it could participate in the next oceanic episode, which are all necessary for the episodic oceanic hypothesis to work, are left unexplained.

7 ACCRETION AND EVOLUTION OF WATER

In the previous chapters we have built a case for extensive modification of the surface by water. In this chapter we discuss what the total inventory of water near the surface of the planet might be and how it may have evolved and been redistributed with time. Water is only one volatile species. Its fate is likely to have been coupled to the fate of other volatiles, such as CO_2, N_2, and the noble gases, so these are also briefly discussed. Up to this point the book has been mainly concerned with surface morphology because it has provided most of the evidence that Mars is water rich and that water has played a prominent role in the evolution of the planet. Indeed, had the surface been masked from the view of spacecraft and only the atmosphere been accessible to measurements, water would not have been an issue, at least until recently. But quantitative measures of the amount of water from geology alone are difficult to obtain, and in this chapter we examine some of the geochemical evidence for the evolution of water.

The present inventory and composition of volatiles on the planet is the cumulative result of a variety of processes that have acted on the planet during both its formation and its subsequent evolution (Figure 7-1). The amount of water retained by the planet at the end of accretion must have depended on the water content of the materials that accreted to form the planet and on the ability of the planet to retain the water during accretion. Part of the water retained would have been within the solid planet, part would have been at the surface, and part would have been in the primitive atmosphere. Some of that within the solid body would have been expelled from the interior during global differentiation at the end of accretion, and some would have been outgassed subsequently as a result of volcanic activity. Water in the atmosphere at the end of accretion could have met a variety of fates, such as precipitation onto the surface, and loss to space

as a consequence of large impacts and a number of processes operating at the top of the atmosphere, stimulated by solar radiation and interaction with the solar wind. *Hydrodynamic escape*, a process whereby loss of hydrogen from the top of the atmosphere entrains heavier molecules, which are also lost, is thought to have been particularly effective at the end of accretion.

After accretion, during heavy bombardment, a change probably occurred in the nature of the bodies impacting the planet, as those in near-circular orbits were rapidly swept up by the growing planets, leaving those in more eccentric orbits. With this change the mix of materials impacting the planets may have changed, and the mean impacting velocities probably increased, although this is by no means certain. According to some models, the Earth acquired a volatile rich veneer during this stage, and Mars may also have acquired one. Global fractionation and core formation at the end of accretion must have resulted in high heat flows and massive outgassing of the interior. High heat flows and high rates of volcanism probably continued throughout the period of heavy bombardment. The volcanism, combined with high impact rates, resulted in efficient mixing of at least the upper several kilometers of the planet. Climatic conditions during heavy bombardment are unknown, but high erosion rates at the end of heavy bombardment suggest warmer conditions. If so, then carbonate formation may have been efficient, and any carbonates that formed may have been mixed into the shallow interior. Losses from the top of the atmosphere would have continued, although hydrodynamic escape would have declined substantially. If the climate was warmer and more water was present in the atmosphere, loss rates of hydrogen, and hence water, from the top of the atmosphere may have been significantly larger than during the rest of the planet's history.

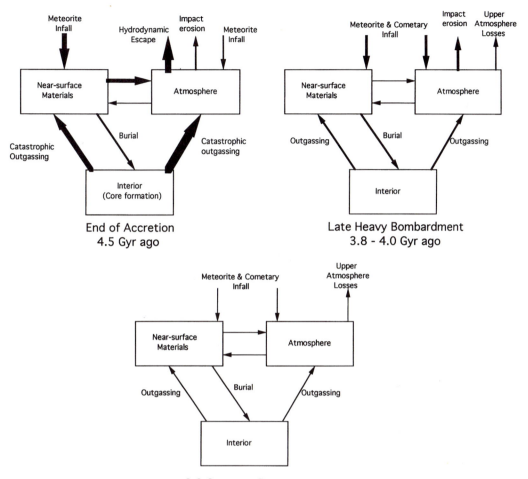

Figure 7-1. Exchange processes affecting the acquisition, loss, and redistribution of water (a) at the end of accretion, (b) at the end of heavy bombardment, and (c) from the end of heavy bombardment to the present.

After heavy bombardment rates of exchange between the various sources and sinks slowed considerably. The lower impact rate resulted in trivial additions to the volatile inventory from meteorite and cometary infall, and trivial losses by impact erosion. The interior continued to outgas, as evidenced by volcanic materials at the surface and detection of it in the atmosphere (Krasnopolsky et al., 1994), but outgassing was probably at a declining rate. Losses from the upper atmosphere would have continued to deplete the near-surface inventory. Despite these rather modest additions and losses to the near-surface reservoirs, water continued to be redistributed between the atmosphere and near-surface materials, and among the near surface materials, to form the various erosion features that we have discussed.

Inventories of volatiles, more specifically H_2O and CO_2, have been estimated in a variety of ways. Geologic estimates of the amount of water at the surface are based on the record of erosion left on the surface and the amount of outgassing, as inferred from the volume of volcanic rocks. They thus concern only the amounts present in the near-surface and atmosphere boxes, in Figure 7-1 c, at various times since 3.8 Gyr ago. Other estimates are based on the present composition of the atmosphere, and how it might have changed with time as a result of exchanges with the surface, losses from the upper atmosphere and additions through outgassing. These estimates also allow projection backward only to the end of heavy bombardment because different processes were probably operating before that time. Other estimates start by trying to model what might have happened during accretion and then step forward and try to match what we know of the present chemistry of the atmosphere and mantle. Only recently have comprehensive models been devised that try to take into account all the various transfers of volatiles shown in Figure 7-1 and how their rates may have changed with time (Jakosky et al., 1994; Pepin, 1994).

Geochemical and Geophysical Processes Affecting the Acquisition and Redistribution of Water

Before discussing how much water might have been acquired by Mars and what its fate might have been, it will be useful to review the nature of some of the processes involved and some general principles of geochemical reasoning.

Chondritic meteorites

The terrestrial planets accreted from materials condensed from the solar nebula. We appear to have samples of these materials in a class of meteorites called *chondrites* (for an overview of meteorites and their classification, see Sears and Dodd, 1988; Wood, 1979). Ordinary chondrites constitute over 70% of known meteorites. They consist almost entirely of high-temperature (500–1000°C) minerals: olivine, orthopyroxene, Ni-Fe metal, and minor amounts of plagioclase, diopside, and chromite. The most characteristic attribute of chondritic meteorites is the presence of chondrules, which are spherical droplets, mostly 0.01–0.1 mm in diameter. They are thought to have formed by poorly understood, high-energy events that involved melting of pre-existing solids in the dusty nebula (Grossman, 1988).

In ordinary chondrites the chondrules and the matrix are chemically and mineralogically similar, both being high-temperature assemblages. About 5% of all meteorites are carbonaceous chondrites. Like the ordinary chondrites, they have chondrules with high-temperature mineral assemblages, but the chondrules are embedded in a fine-grained matrix of low-temperature, mainly hydrated, minerals such as phyllosilicates, together with lesser amounts of carbonates, sulfates, and Fe-Ni sulfide. The matrix also contains about 5% organic compounds. Carbonaceous chondrites contain, in addition, anhydrous inclusions of refractory materials that are thought to have condensed directly from a vapor. The inclusions are composed of very high-temperature minerals such as spinel, perovskite, and some garnets. Carbonaceous chondrites are thus disequilibrium assemblages, with a high-temperature reduced component embedded in a low-temperature more oxidized component.

Different types of carbonaceous chondrites are recognized, mainly on the basis of the proportions of volatile to refractory components. C1 carbonaceous chondrites are particularly important in accretionary models because almost all elements are present in the same proportions as in the Sun's photosphere (Anders and Grevesse, 1989). They are regarded, therefore, as the best representative of unfractionated material from the primitive solar nebula (Figure 7-2) and are used as a reference to determine fractionation with respect to nebular material. All chondrites have ages of 4.55 Gyr so that this is taken as the age that most of the material condensed from the solar nebula.

Chemical fractionation

As materials were condensing from the solar nebula and the grains were aggregating into larger bodies, the condensed phases were fractionated against uncondensed phases. Chemical differences between the different carbonaceous chondrites result, therefore, largely because of differences in proportions of volatile and nonvolatile components. The temperature at which a given element condenses from a gas of nebular composition with a pressure of 10^{-4} bars is generally used as a measure of volatility (Table 7-1). Differences in the bulk composition of terrestrial planets can similarly be modeled as mixes of materials of different volatilities, on the assumption that during accretion elements of like volatility were incorporated into the planets in like proportions with respect to the solar nebula composition, as represented by C1 chondrites.

Once materials became incorporated into planets, however, volatility was no longer an important factor in fractionation, except for the highly volatile components, such as H_2O and CO_2 and the noble gases. Fractionation within hot planetary interiors would have depended more on chemical affinity and mineralogic compatibility. Three main chemical affinities are recognized. In

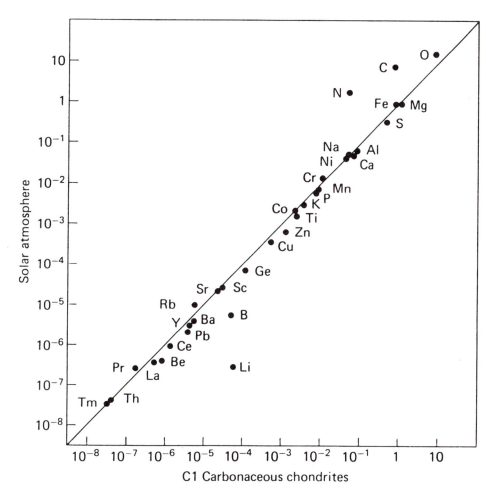

Figure 7-2. Comparison of elemental abundances in Type 1 carbonaceous chondrites with those in the solar photosphere. Abundances are normalized to Si = 10^6. (From Ahrens et al., 1989. Reproduced with permission from the University of Arizona Press.)

coexisting silicate, metal, and sulfide melts, lithophile elements (e.g., Mg, Al, U, Th) preferentially partition into the silicate melts, siderophile elements (e.g., Ni, Co, Ir) preferentially partition into the metal melts, and chalcophile elements (e.g., As, Cu, Sb) preferentially partition into the sulfide melts (Figure 7-3). Thus formation of the Earth's core preferentially scavenged siderophile elements from the mantle to leave it depleted in siderophiles with respect to C1. The martian core is believed to contain significant fractions of Fe-Ni sulfide on the basis of chemical and geophysical data (Goettel, 1981; Wänke, 1981), so formation of the core left the martian mantle depleted in chalcophiles (Figure 7-4).

Compatibility is a measure of the ability of a particular element to be accommodated in the major mineral phases in the mantle. In partial melting, incompatible elements preferentially enter the melts and during crystallization they tend to concentrate in the residual melts. On Earth, incompatible elements, such as U, Th, and K, tend to be preferentially concentrated in the crust. Although not an element, water acts as if it were incompatible, remaining largely in residual liquids in a crystallizing basalt, for example.

Hydrodynamic escape

Hydrodynamic escape is a mechanism recently invoked, primarily to explain the observed fractionation patterns of noble gases in the atmospheres of Venus, Earth, and Mars (Hunten et al., 1987, 1988, 1989; Pepin, 1991; Sasaki and

Table 7.1. Condensation temperatures and sequence of phases and elements separating from gas of Solar composition at 10^{-4} atm. total pressure (From Ringwood, 1979).

Element or Compound	°K
Ca, Al, Ti oxides and silicates	>1400
Platinum metals, W, Mo, Ta, Zr, REE	>1400
Mg_2SiO_4	~1360
Fe-Ni metal	~1360
Remaining SiO_2 (as $MgSiO_3$)	1200-1350
Cr_2O_3	—
P	1290
Au	1230
Li	1225
Mn_2SiO_4	1190
As	1135
Cu	1118
Ga	1075
(K, Na)$AlSiO_3$	~1000
Ag	952
Sb	910
F	855
Ge	812
Sn	720
Zn	660-760
Se	684
Te	680
Cd	680
S (as FeS)	648
FeO (10% ss. $[Mg_{0.9}Fe_{0.1}]_2SiO_4$)	~600
Pb	520
Bi	470
In	460
Tl	440
Fe_3O_4	400
NiO	—
H_2O (as hydrated silicates)	~300

GEOCHEMICAL GROUPING OF ELEMENTS

	Refractory	Moderately volatile	Volatile
Lithophile Elements	Al, Si, Mg, Ca, Sc, Hf Lu, Ta, Ti, Th, La, Eu, V, U, Ba , Cr, Sr	Li, Na, K, Ga, Rb	Cl, F, Br, Zn, Cs
Chalcophile Elements		As, Cu, Ag	Sb, Se, S, In, Bi, Tl
Siderophile Elements	Re, W, Ni, Co, Os, Ir, Mo	Au, P, Mn, Ge	

Figure 7-3. Geochemical grouping of elements. Geochemical affinity is not an absolute property of an element. Partitioning between different melts depends on the specific geochemical and physical conditions. Groupings shown here are for typical mantle conditions.

Nakazawa, 1988; Zahnle and Kasting, 1986; Zahnle et al., 1990). It was effective only very early in the history of the solar system, when hydrogen-rich primitive atmospheres, thought to have been present on young planets, were heated at high altitudes by high fluxes of extreme ultraviolet radiation (EUV) from the young sun. Fluxes of hydrogen, thermally escaping from the upper atmosphere, were high enough to create, in effect, an outwardly directed hydrogen wind. The drag forces exerted by the wind on heavier atmospheric constituents, combined with the high temperatures, were sufficient to lift them out of the atmosphere. The rates of escape of different species depend on their mole fraction and molecular weight. Light species were lost more readily than heavier species so mass fractionation resulted.

The mechanism is one of the few known that can explain the observed fractionation of very heavy species such as Xe, for which, in the atmospheres of both Earth and Mars, the different isotopes are fractionated both with respect to each other and with respect to solar and meteoritic Xe. For the process to work for elements as heavy as Xe, the planet's atmosphere must be very rich in hydrogen and the EUV flux from the Sun must be two to three orders of magnitude greater than the present flux (Pepin, 1991). Observations of soft x-ray luminosities of solar-type, pre-main sequence stars suggest that such high fluxes levels could have been reached by the

early Sun (Walter et al., 1988; Fiegelson et al., 1989). The decline in soft x-ray luminosities with stellar age indicates that the early high EUV flux declined with a decay constant of roughly 90 Myr (Pepin, 1991), but hydrodynamic escape may have been ineffective during the first 50 Myr because of a nebular opacity caused by dust and gas. The efficacy of the mechanism clearly depends on the availability of abundant hydrogen. Pepin (1991) envisaged a two-stage process. In the first stage, starting at a solar age of about 50 Myr, hydrogen in the primordial atmosphere, present at the end of accretion, is rapidly exhausted. In the second stage, starting at a solar age of about 150 Myr, hydrogen outgassed from the interior as a result of core formation rekindles hydrodynamic escape, but it is less efficient at this second stage because of the decline in the EUV flux. Subsequent modeling (Pepin 1994) suggests, however, that a two-stage process may not be necessary because of sputtering effects. Hydrodynamic escape is effectively over after about 350 Myr or 4.2 Gyr ago.

Impact erosion

Impact erosion of the atmosphere was discussed briefly in the previous chapter. Because of the smaller radius and escape velocity for Mars, its atmosphere is more vulnerable to blow-off by large impacts than are the atmospheres of Earth and Venus. For impact erosion to occur, most of

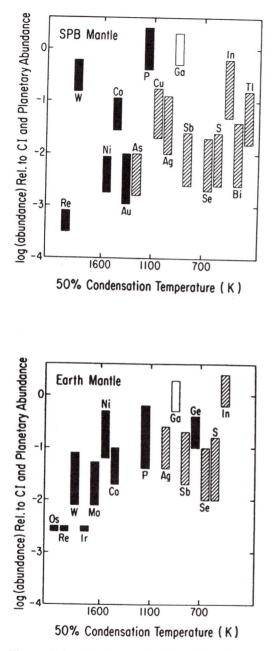

Figure 7-4. Abundance of siderophile elements (black bars) and chalcophile elements (hachured bars) normalized to C1 carbonaceous and lithophile elements of similar volatility. The upper diagram is for the martian mantle (Shergotty Parent Body); lower diagram is for the Earth's mantle. Most chalcophile and siderophile elements are more depleted in the martian mantle. The relatively high W and P in the martian mantle may result because the martian mantle is more oxidized. (From Treiman et al., 1986). Reproduced by permission from Elsevier Science.)

the vapor plume that forms at impact must reach escape velocity. This requires a minimum impact velocity of 14.3 km s^{-1} for silicate impactors and 11.1 km s^{-1} for ice impactors (Melosh and Vickery, 1989). Root mean square impact velocities at Mars are 21 km s^{-1} for periodic comets, 42 km s^{-1} for parabolic comets, 19 km s^{-1} for Earth-crossing asteroids, and 10 km s^{-1} for Mars-crossing asteroids, so that most impactors are likely to exceed the minimum velocity, at least for present-day Mars. Impact erosion also requires that the mass of the plume exceed the mass of the atmosphere above a plane tangent to the target, so erosion depends on atmospheric pressure and impactor size. The efficacy of impact erosion on early Mars depends, therefore, on the impact rate, the mix of impacting projectiles, and the thickness of the atmosphere. Impact erosion is non-fractionating against the various components in the atmosphere but should have been more efficient for the noble gases as opposed to volatiles such as water that could condense on the surface or volatiles such as CO_2 that could react with the surface.

Outgassing

Whole rock U-Pb, Pb-Pb, and Rb-Sr data all indicate that the SNC meteorites are derived from a common reservoir in the mantle with a model age of close to 4.5 Gyr (Jones, 1986; Shih et al., 1982; Jagoutz, 1991). The age is essentially indistinguishable from the age of the solar system as derived from chondritic meteorites, so global differentiation and core formation probably took place at the end of accretion. Core formation would have resulted in reaction between the varous components that accreted to form the planet and in catastrophic outgassing of volatiles. As a consequence of accretional heating and core formation, the mantle was almost certainly hot and convecting vigorously during the first few 100 Myr of the planet's history, and the heat flow was high and rapidly declining. Thermal modeling indicates that the heat flow probably declined from several hundred milliwatts per square meter (mW m^{-2}) at the time of core formation to around 150 mW m^{-2} at the end of heavy bombardment, after which time it declined slowly (see Figure 1-14) to its present estimated value of 30 mW m^{-2} (Stevenson et al., 1983; Schubert and Spohn, 1990). Volcanism, and hence out-gassing, is likely to have followed a similar pattern, declining rapidly for a few 100 Myr, then declining slowly for the rest of Mars' history (Schubert et al., 1992). Estimates of the volumes of volcanic rocks, although reliable only from the beginning of the Hesperian, which probably dates back to 3.5–3.8 Gyr ago, appear to show a declining rate of volcanism, as expected from the thermal modeling (Greeley, 1987; Greeley and Schneid, 1991; Tanaka et al., 1988).

Upper atmosphere losses

Atmospheric components can be lost from the upper atmosphere by a variety of processes. Photodissociation of atmospheric components by ultraviolet radiation in the lower atmosphere and recombination of the dissociation products result in a variety of molecular, ionic, and atomic species, which are transported up to the exobase at about 200 km altitude. Above the exobase, collisions are infrequent and atoms and molecules are on ballistic trajectories and thus are vulnerable to escape (McElroy et al., 1977; Hunten, 1993). Thermal or Jeans escape results when atoms in the high-velocity tail of the Maxwellian distribution are directed upward and exceed the escape velocity of 5.03 km s^{-1}. The efficiency is strongly mass dependent and significant only for hydrogen and deuterium (Yung et al., 1988). Thermal escape has resulted in significant enrichment of the atmosphere in deuterium because of enrichment of the exosphere in hydrogen as a result of gravitational separation of hydrogen and deuterium above the homopause. Heavier species may be lost as a result of dissociative recombination, which is a process whereby a molecular ion, such as N_2^+ or O_2^+, combines with an electron to produce two fast atoms, which may have sufficient energy to escape. Again, the process favors the lighter isotope because of gravitational separation.

Atmospheric sputtering causes additional losses. Neutrals in the upper atmosphere become ionized and then "picked up" by magnetic fields generated by flow of the solar wind around the planet (Zhang et al., 1993). These pick-up ions spiral around the magnetic field lines as they are carried around the planet into the magnetotail. Some of the ions impact the exobase and, by mo-mentum transfer collisions, give atmospheric atoms or molecules sufficient energy to escape (Luhman and Kozyra, 1991). Direct sputtering by hydrogen and helium ions appears to be negligible compared with sputtering by pick-up ions. The total loss of C from CO_2 by sputtering over the last 3.5 Gyr was estimated by Luhmann et al. (1992) to be equivalent to 120 mbar CO_2, and the amount of oxygen lost by sputtering and other processes was estimated by Luhman et al. to be equivalent to a 30-m thick layer of water. Because of the lack of substantial fractionation of $^{18}O/^{16}O$ or $^{13}C/^{12}C$, there must be nonatmospheric O and C reservoirs, at least equivalent in size to the amounts lost (Jakosky et al., 1995). These numbers are very uncertain because of uncertainties in how the UV flux has changed over the last few giga-years, and in the yields from dissociative recombination and sputtering. As before, all the sputtering reactions favor the lighter isotopes because of gravitational separation above the homopause.

Sputtering can readily explain the observed ratios of $^{36}Ar/^{38}Ar$ and $^{22}Ne/^{20}Ne$. There is no need to invoke a second period of hydrodynamic escape to specifically explain the fractionation (see earlier section on hydrodynamic escape). Jakosky et al. estimate that 50 to >90% of all the Ar supplied to the atmosphere has been lost by sputtering. Ne is lost so readily that supply by outgassing is required. This resupply appears confirmed by detection of He in the martian atmosphere, where it has a lifetime of only about 50,000 years (Krasnopolsky et al., 1994).

Accretion

The terrestrial planets must have acquired their initial atmospheres and inventory of volatiles either by inheriting them from the nebula or by extracting the volatiles in some way from the solid materials that accumulated to form the planets. Most current theories largely discount acquisition of a nebular contribution, mainly because of the scarcity of noble gases, but with recent recognition of the efficacy of hydrodynamic escape and impact blow-off in removing noble gases, this conclusion should be questioned. Nevertheless, in the following discussion contributions from nebular gases are assumed to be insignificant. In recent years a "standard model" of planetary accretion has evolved. In the gaseous, turbulent and dusty nebular disc, dust grains are thought to have readily agglomerated into 0.1–1 cm objects. These agglomerations then gravitationally settled to the midplain of the nebula disc (Weidenschilling, 1984). The grains included both chondrules and the very refractory assemblages found as inclusions within carbonaceous chondrites. In the midplane these agglomerations accumulated into 1–10 km objects, termed *planetesimals*, by processes that are very poorly understood. At this early stage the nebula gas kept the solid materials in roughly circular orbits and moderated collisional velocities. Once the planetesimals formed the nebula gas is thought to have dissipated, and subsequent evolution of the planetesimal swarm was dominated by gravitational interactions between the planetesimals, (Safranov 1972; Wetherill and Stewart, 1989). The first stage was accumulation of the more than 10^{10} roughly 10 km objects required to make the terrestrial planets into a relatively small number (~30) of planetary embryos. Modeling suggests that this took place rapidly, within 10^5 to 10^6 years (Wetherill and Stewart, 1989). The final stage was accumulation of the present planets from these embryos, which is estimated to have been accomplished in 10^7 to 10^8 years (Wetherill, 1990).

As the planets grew in size, impact velocities would have increased. Ultimately impact velocities became so large that volatiles in the impacting objects, which were presumably of varying chondritic compositions, would have become vaporized and an atmosphere would have started to form (Benlow and Meadows, 1977; Jakosky and Ahrens, 1979; Gerasimov and Mukhin, 1979). Thus, in the early stages of planetary growth most of the volatiles in the accreting objects were probably incorporated into the growing solid planet, but when the threshold size was reached for impact devolatilization, a significant fraction of the volatiles was driven off at the surface. Water is by far the most common volatile in meteorites, so the main concern is with water. The extent to which water was incorporated into the planet after the devolatilization threshold was reached would have depended on the extent to which rehydration of the silicates took place at the surface. This would have been favored by temperatures sufficiently high for liquid water to occur but below dissociation temperatures for the hydrated minerals.

Abe and Matsui (1985, 1986), Matsui and Abe

(1987), and Zahnle et al. (1988) modeled the accretion of the terrestrial planets on the basis of the scenario just outlined. The Abe and Matsui modeling indicates that in the case of the Earth (and Venus), the water released at the surface formed a massive steam atmosphere that impeded loss of heat from the surface. They assumed that when surface temperatures were below 900°K a large fraction of the water in the accreted material was consumed in rehydrating the silicate crust, but when temperatures exceeded 900° C almost all the water released entered the atmosphere. Because of the thermal blanketing by the steam atmosphere, the accretional energy could not be radiated away and surface temperatures rose. Ultimately the surface melted. According to the model, the Earth would have started developing a steam atmosphere at about 0.2R and the surface would have melted at about 0.4R, where R is the final radius of the Earth. Once the surface melted, the atmosphere was maintained at roughly 100 bars surface pressure by water dissolving in the surface melt. Water continued to dissolve in the melt to maintain a constant surface pressure as more water was added to the planet with each impact. At the end of accretion, when the impact rate fell, solar and accretional energies were insufficient to maintain the high surface temperatures, the steam condensed and the oceans formed. Thus, the Earth was left with abundant water on the surface (Figure 7-5).

The modeling suggests that the accretion of Mars would have been very different (Matsui and Abe, 1987). Whether a steam atmosphere developed and the surface melted would have depended on the accretion time. The standard model of Abe and Matsui assumes an accretion time of 5×10^7 years, within the range suggested by Wetherill (1990). However, for the surface of Mars to melt in their model, accretion times of 5×10^6 years or shorter are required (Figure 7-6). For accretion times within the range suggested by Wetherill, a steam atmosphere never forms on Mars and the surface never melts. Impact devolatilization would have resulted in release of water to the surface, starting at about 0.4R, but the water would have condensed as water or ice. Thus in the late stages of accretion Mars may have accumulated a water-rich surface layer, but the process of incorporation of water deep into the interior may have been very different on Earth and Mars. For Earth the process was solu-

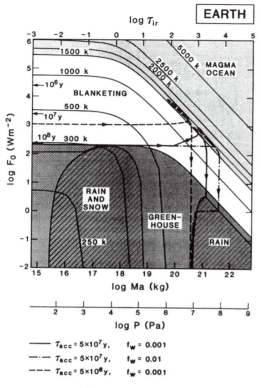

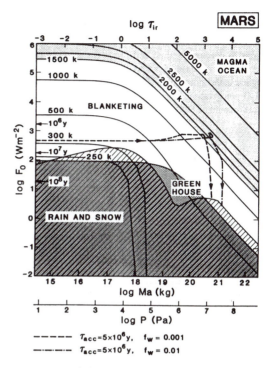

Figure 7-6. Same as the previous figure but for Mars. For the more probable accretion times of greater than 10^7 years, the evolutionary tracks enter the liquid water region, almost all the water in the atmosphere condenses and so a steam atmosphere does not develop, although the planet is left with a water-rich surface. A steam atmosphere develops only for accretion times shorter than 10^7 years, which are improbable according to current models. (Matsui and Abe, 1987, as modified by Ahrens et al., 1989. Reproduced with permission from the University of Arizona Press.)

Figure 7-5. Evolutionary tracks for surface temperatures for the Earth during accretion, plotted as a function of the energy flux F_0 released at the surface, and the mass Ma, pressure P, and optical depth τ_{ir} of an H_2O atmosphere. The light solid numbered curves are surface temperatures. The shaded region is where the solar flux dominates the thermal structure of the atmosphere. The unshaded area is where, the surface is blanketed and energy released at the surface dominates F_0. The stippled area shows where a magma ocean forms. In the cross-hatched regions, precipitation occurs. The arrows on the y-axis show the accretion times. The thick solid line, starting close to 2 Wm^{-2} on the y-axis, shows the sequence of events for an accretion time of 10^8 years. As the steam atmosphere builds, and surface temperatures rise, the evolutionary track moves from the shaded area to the unshaded until the surface melts. At this point part of the atmosphere dissolves in the surface melts. As accretion continues the atmosphere is maintained at a roughly constant pressure by solution of water in the surface melts. When accretion slows, F_0 declines. The track then moves vertically down the diagram until it intersects the cross-hatched region, at which point the atmosphere precipitates out to form the oceans. (From Matsui and Abe, 1987, as modified by Ahrens et al., 1989. Reproduced by permission from the University of Arizona Press.)

tion from a massive atmosphere; for Mars the process would have been the less efficient process of burial and foundering of a hydrated silicate crust.

Zahnle et al. (1988) elaborated upon the Matsui and Abe modeling for formation of the Earth by incorporating a number of processes that were not included in the original model. The two most important additions were inclusion of hydrodynamic escape and impact erosion of the atmosphere. They found that their models agreed with the Matsui and Abe models for a wide range of conditions, but other results are possible. If the water content of the impacting planetesimal was low and hydrodynamic escape was sufficiently high, then a steam atmosphere would not develop. Thus, if the Earth accreted inhomogeneously, with the volatile-rich materials arriving late as suggested by Wänke (1981) and Newsom (1990), then development of the steam atmosphere may have been delayed until the later stages of accretion. The presence of a steam atmosphere also means that some significant fraction of the water in the atmosphere could be lost by impact erosion. Zahnle et al. also point out that if the planetary surface does not melt, as may have occurred with Mars, then incorporation of water into the interior would have depended on the efficiency of the hydration and foundering of the surface materials. If hydration required the presence of liquid water, it could have been very inefficient, allowing little of the water that accumulated at the surface to get into the interior. Thus toward the end of accretion Mars could have had a water-rich surface. The vulnerability of this water to loss by hydrodynamic escape and impact erosion would have depended on the surface conditions and how the water was distributed between the atmosphere and the surface, about which we have no information.

Water Retention After Core Formation

One of the few events during this early period for which we have unequivocal evidence of its timing is core formation. As indicated earlier, isotope systematics show that the core segregated 4.5 Gyr ago, at the end of accretion (Shih et al., 1982; Chen and Wasserburg, 1986; Jagoutz, 1991). The Earth's core also formed at this time, but the subsequent history of the interior of the two planets is very different. Long after core formation, mixing by convection continued within the Earth's interior so that now the Earth's mantle has an average model age of 2 Gyr. In contrast, the Mars mantle appears to have remained unmixed with a model age of 4.5 Gyr (Jagoutz, 1991). If it has been convecting, it has been doing so in a way that has not caused fractionation between radiogenic species and their products.

Mars is likely to have acquired large amounts of water within its interior during accretion, despite the difficulty of incorporating surface water into the interior, because it grew to 0.4 times its present size before onset of impact devolatilization. Dreibus and Wänke (1987) estimated the amount of water incorporated into the planet by first estimating the composition of the mantle, then determining what mix of meteorites best fit this composition. They estimated the composi-

tion of the mantle from the chemistry of SNC meteorites on the assumption that refractory elements were accreted in proportion to their contents in chondrites, but that the other elements were incorporated in lesser proportions depending on their volatility. Mg and Fe are major mantle components. Mg, being highly refractory, was assumed to be present in the same proportion to Si as in C1 carbonaceous chondrites. Mn is also present in SNC meteorites in the same proportion to Si as it is in C1 carbonaceous chondrites and so was assumed to be unfractionated with respect to Si during planet formation. Since Mn and Fe are barely fractionated against each other in most magmatic processes, the Fe/Mn ratio in SNCs gives the mantle Fe/Mn content, which, combined with the inferred Mn content, gives the Fe abundance. They noted that the abundances of several key volatile elements, such as K and Br, have a strong correlation with the abundances of La, a refractory element. The ratios of these volatiles to La thus gives a depletion factor for other elements of comparable volatility (Figure 7-7). In this way they were able to reconstruct the

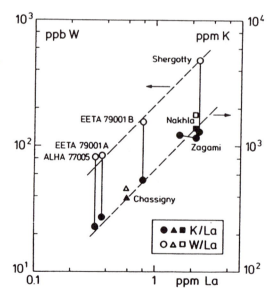

Figure 7-7. Correlation of La with K (solid markers) and W (open markers) in SNC meteorites. The correlations indicate that these elements follow each other in magmatic processes and are not fractionated with respect to each other. The constant K/La ratio can be used to estimate the ratio of moderately volatile elements, represented by K, to refractory elements, represented by La, that were incorporated into the planet. (Reproduced by permission from Dreibus and Wänke, 1984.)

chemistry of the martian mantle, and their composition agrees well with the mantle density as inferred from geophysical measurements (Goettel, 1981). The composition of the Earth's mantle has also been determined, but from analyses of primitive, uncontaminated spinel-lherzolite xenoliths (Jagoutz et al., 1979) thought to be mantle samples. Dreibus and Wänke found that many moderately volatile elements, such as Fe, Ga, Na, K, P, F, and Rb, are more abundant in Mars' mantle than in the Earth's mantle and thus they concluded that Mars is more volatile rich than the Earth.

Having inferred the composition of the mantle, Dreibus and Wänke (1987) estimated the amount of water incorporated into Mars during accretion on the basis of a two-component accretionary model proposed by Ringwood (1977, 1979) and later modified by Wänke (1981). According to this model, the terrestrial planets accreted from a mix of materials that can be represented by two end member components. Component A is highly reduced, Fe and all siderophile elements are present as metals, and elements with volatilities of Na or higher are absent. In component B all metals are oxidized, and all elements, including the volatiles, are in the proportions found in C1 carbonaceous chondrites. Dreibus and Wänke showed that the chemistry of the Earth's mantle is well matched by a mix of 85% component A and 15% component B. The chemistry of the Mars mantle is well represented by a mixture of 60% component A and 40% component B. Thus Mars incorporated a larger fraction of the volatile-rich component than the Earth, as expected from its greater distance from the Sun.

During core formation the interior would have melted and equilibrated. Any water present in the interior would have reacted with metallic iron to produce FeO and hydrogen, which would have outgassed. The Earth's mantle contains higher proportions of siderophile elements (Ni, Co, W, etc.) than expected if it equilibrated with the core, as it should have done when the planet underwent global differentiation near the end of accretion. To explain this anomaly, Dreibus and Wänke suggested that some of component B was added to the Earth late, after core formation, so never equilibrated with the core. Because of the continued late addition of oxidized, water-rich material late in accretion, the Earth's mantle changed from a reduced to a more oxidized condition until metallic iron was no longer stable. The small fraction of component added late, after metallic iron was no longer present in the mantle, has been termed a *late volatile-rich veneer* (Wänke, 1981; Newsom, 1990), and may be the source of much of the water on the surface of the Earth. (This suggestion is not necessarily in conflict with the steam atmosphere models described earlier because the late veneer could have contributed to the steam atmosphere.) The 3.2 ppb of the highly siderophilic Ir in the Earth's mantle (Jagoutz et al., 1979) indicates that the late veneer, added after core formation, constitutes only 0.44% of the Earth's mass. We should note in passing that acquisition of a late volatile-rich veneer is only one possible interpretation of the siderophile anomaly. For alternatives see Brett (1984), Newsom (1990), and Murthy (1991).

The Mars mantle appears not to have a siderophile anomaly. There are ambiguities because of volatility effects, and because chemical affinity is not a fixed elemental property but depends on the specific conditions during partitioning (Treiman et al., 1986). The Mars mantle is more depleted in chalcophile elements than is the Earth mantle. Estimates of density of the core and depletion of chalcophiles indicate that the core is sulfur rich (Goettel, 1981; Trieman et al., 1986). Sulfur is mostly derived from component B. To explain the lack of a siderophile anomaly and the presence of sulfur in the core, Dreibus and Wänke suggested that, in contrast to the Earth, components A and B accreted together to form Mars. As a consequence, all the water in component B reacted with the metallic iron in component A to form FeO and hydrogen, which escaped. The massive loss of hydrogen from the interior, they suggested, would have removed any surface atmosphere by hydrodynamic escape. They concluded, therefore, that at the end of accretion the only water retained by Mars was that dissolved in the mantle. From solubility relations and comparisons with other volatile elements such as Cl and Br, they estimated that after core formation the water content of the mantle was 36 ppm, which is equivalent to a global layer 130 m deep.

If we assume the Scambos and Jakosky (1990) factor of 0.017–0.12 for subsequent release of interior volatiles to the surface, then 2–16 m of water would have been released to the surface. If we take the upper limit of 0.5 suggested by Jakosky et al. (1995) for the release factor, then as much as 65 m could have been released to the surface. Thus the Dreibus and Wänke model implies a very dry Mars, much drier than geologic estimates based on water erosion (see later). Some support for the dry interior postulated by Dreibus and Wänke is the low water content of SNC meteorites, which mostly range from 130 to 350 ppm as compared with 1500–6000 ppm for terrestrial oceanic basalts (for a summary, see Carr and Wänke, l992). The low water contents may, however, simply reflect efficient outgassing as the basalts are brought to the martian surface.

Although Dreibus and Wänke suggest a very dry Mars after formation of the core, there are several possibilities, consistent with the geochemical evidence, whereby Mars could have retained substantial amounts of water both in the interior and on the surface. Reaction of water with metallic iron may not have been 100% efficient. Dreibus and Wänke estimated that during accretion Mars incorporated 3.36% H_2O, or roughly the equivalent of 150 km spread over the whole surface. Thus, if only 2% did not react with metallic iron, the equivalent of 3 km was left in the interior to be subsequently outgassed. This is ample to explain the geology. Another possibility is that Mars acquired a late volatile-rich veneer as has been proposed for the Earth, but the veneer left no signature on mantle chemistry because of the lack of plate tectonics (Carr and Wänke, l992). Indeed, it is difficult to see how the Earth could have acquired such a veneer without Mars doing so also. Roughly 1 km of water would be added to the planet if the outer 5 km (0.4% of the planet's mass) was added late, so only a modest fraction of component B in the Dreibus and Wänke model needed be added after core formation to provide enough water to explain the geology. Another possibility is that substantial amounts of water were left on the surface at the end of accretion, as suggested by the Abe and Matsui models. If core formation did not result in surface melting and if surface temperatures were such that the water was condensed on the surface, then a significant fraction of this could be retained.

Finally, Chyba (l990) suggested that the Earth could have acquired its oceans as a consequence of accretion of a veneer of material from comets and asteroids during heavy bombardment between 4.5 and 3.5 Gyr ago. He took the lunar cratering record, appropriately scaled for the Earth's larger cross section, and found that the Earth had acquired between 10^{21} and 7 x 10^{23} kg of material

during this period as compared with 1.4 x 10^{21} kg for the mass of the oceans. Taking into account ejection of condensed water from the oceans and loss of water from the accreting bodies as a result of escape in a vapor plume during impact, he estimated that the Earth had acquired and retained 0.2–0.7 ocean masses of water as a result of impacts during heavy bombardment. Because of the lower escape velocity on Mars, ejecta is less efficiently retained, and he estimated that Mars would have acquired 10–100 m of water during this same period. Such estimates must involve large errors, depending as they do on knowing the mix of comets and asteroids, the thickness of the atmosphere, and the impact velocities, and being able to model large impacts. Some support for late acquisition of volatiles is provided by the noble gases. Owen et al. (l992) showed that on a plot of $^{36}Ar/^{132}Xe$ against $^{84}Kr/^{132}Xe$ the atmospheres of all of the terrestrial planets lie on a mixing line between two components. One component that they termed an internal component is assumed to have outgassed from the interiors. The other component has noble gas ratios similar to experimentally determined ratios found when noble gases are adsorbed onto low temperature ices. They suggested that this latter component was cometary and that contributions from comets are significant in the present martian atmosphere.

We can conclude from these two sections that there are no compelling reasons why Mars should not have emerged from accretion and core formation without a substantial inventory of water. Indeed, acquisition of a larger initial fraction of volatiles and the colder surface temperatures at the end of accretion may have more favored water retention on Mars than on Earth. On the other hand, if the surface melted when the core formed and water at the surface formed a steam atmosphere, then most could have been lost by impact erosion and hydrodynamic escape. In addition, most of the internal water could have been lost by reaction with metallic iron. We will have to look to the planet's subsequent history to see which of these scenarios is most likely.

Estimates of H_2O and CO_2 from Argon and Nitrogen in the Atmosphere

Anders and Owen (l977) argued, on the basis of the low abundance of noble gases in the atmosphere, that Mars is poor in volatiles and has outgassed no more than the equivalent of 10 m of water averaged over the whole planet. Their arguments were made before impact erosion of atmospheres and hydrodynamic escape were recognized as significant processes, and before SNC meteorites provided reliable means of estimating global chemistry. They noted that the ratios of noble gases in the atmospheres of Earth and Mars are both similar to each other and to those in carbonaceous chondrites, but that the proportions of noble gases to total mass is two orders of magnitude lower on Mars than on the Earth. Because of the high molecular weight of Ar and its lack of reactivity, they assumed that all the Ar that had outgassed from the planet was still present in the atmosphere.

From the measured ^{40}Ar and ^{36}Ar in the atmosphere and estimates of the K content of the mantle, they then estimated what they called *relative release factors*, which are basically ratios describing the outgassing efficiency for different species on Earth and Mars. They found, for example, that ^{36}Ar had a relative release factor of 0.27, implying that it had less efficiently outgassed on Mars by almost a factor of four with respect to the Earth. To estimate the amount of water on the surface, they took the ^{36}Ar abundance in the atmosphere, as determined by the Viking landers (Owen et al., 1977), assumed the terrestrial ^{36}Ar/H_2O ratio, made appropriate corrections for the fraction released, and derived the value of 10 m of H_2O spread over the whole planet for the amount of water released to the surface since the end of accretion. They similarly estimated that the equivalent of 100 mbar of CO_2 had been outgassed to the surface, and presumed that it was mostly sequestered in the surface materials. Implicit in both these estimates is an assumption that Mars had no H_2O and no CO_2 at the surface at the end of accretion and that the amounts present now are the result of outgassing from the interior.

In retrospect, it is now clear that these estimates are in error and probably significantly underestimate the amount of water and CO_2 present. Mars may have lost significant amounts of its original argon as a result of hydrodynamic escape, impact erosion, and sputtering. Much of the near-surface water may never have been within the planet's interior, and there is little reason to

assume that the ^{36}Ar/H_2O/CO_2 ratios should be the same on Earth and Mars. Nevertheless, use of Ar as an index of outgassing efficiency is a reasonable approach, and knowledge of release factors may prove useful in estimating the amount of H_2O and CO_2 released to the surface subsequent to the end of accretion.

Scambos and Jakosky (1990) were able to make better estimates of release factors for volatiles from the interior by taking advantage of more accurate estimates of the bulk potassium content of the planet from SNC meteorite data (Dreibus and Wänke, 1987) and better understanding of the history of volcanism (Tanaka et al., 1988; Greeley, 1987) that has accrued since Anders and Owen made their original estimates. In SNC meteorites and lunar and terrestrial basalts, the refractory element La correlates well with the moderately volatile element K. In other words, igneous processes do not fractionate La with respect to K, so the K/La ratio remains constant. La is a refractory element, and K is a moderately volatile element. The K/La ratio in the SNCs is 1.6 times that in the Earth's mantle (Dreibus and Wänke, 1987), which probably implies that during accretion Mars incorporated a larger fraction of moderately volatile elements than did the Earth (Figure 7-7), as discussed earlier. The bulk K content of the Earth is estimated to be 147–185 ppm (Ganapathy and Anders, 1974; Wänke, 1981), so the bulk K content of Mars should be in the 235–300 ppm range.

Knowing the present amount of ^{40}Ar in the atmosphere, the production rate of ^{40}Ar from the K content, and how the eruption rate of volcanic rock, and presumably volatiles, varied with time (Figure 7-8), Scambos and Jakosky determined that 0.012–0.040 times the total ^{40}Ar produced had been released to the atmosphere. This release factor is less than that for nonradiogenic volatiles because of the continuing production of ^{40}Ar and the declining rates of volcanism with time. They estimated the release factor for nonradiogenic volatiles, including water, to be 0.17–0.112. Jakosky et al. (1994) subsequently revised these figures upward to account for the loss of ^{40}Ar by sputtering. The release factor for ^{40}Ar could be as high as 0.2 and that for nonradiogenic volatiles as high as 0.5. The revised estimates imply that 20–50% of the water in the interior box in Figure 7-1c has been outgassed to the near surface since

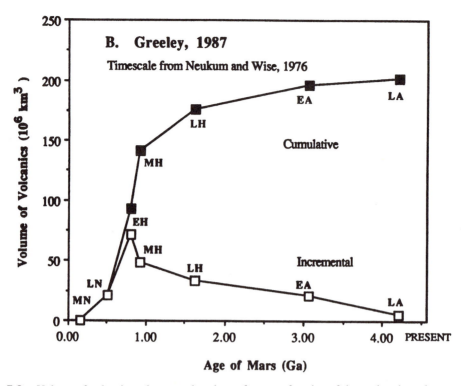

Figure 7-8. Volume of volcanic rocks erupted to the surface as a function of time using the volume estimates of Greeley (1987) and the time scale of Neukum and Wise (1987). The letters on the curves represent time epochs. EH, MH, and LH are early, middle, and late Hesperian, EA and LA, and early and late Amazonian. The figures become more uncertain with increasing age. They are very poorly known for middle and late Noachin (MN and LN) because of the masking effects of high rates of impact and because of burial by later volcanics. (From Scambos and Jakosky, 1990. Copyrighted by the American Geophysical Union.)

the mid-Noachian. It is difficult to see, however, what this means with respect to water inventories because we know neither the water content of the interior nor the amount at the surface.

To estimate H_2O and CO_2 inventories, Anders and Owen (1977) assumed terrestrial ratios with respect to the noble gases. Subsequent discovery that the $^{36}Ar/^{12}C$ ratio in the Venus atmosphere is 20-200 times larger in the volatiles outgassed from the Earth led Pollack and Black (1979) to question that the noble gases and other volatiles were incorporated into the different planets in similar proportions during accretion. They suggested that different mechanisms controlled incorporation of the noble gases and non-noble gases into the grains that accumulated to form the planetesimals from which the planets ultimately formed. Noble gases were mostly adsorbed on the grains, and non-noble gases were mostly chemically bound within the grains. They suggested that prior to formation of the planets, while the nebula was mostly

a dust-gas mixture, the nebula was essentially isothermal, but the nebular pressure decreased outward. Because of the dependence of adsorption on the local nebular pressure, the grains from which Venus accreted had more adsorbed noble gases than those from which Earth and Mars accreted. In contrast, the abundances of CO_2, N_2, and H_2O were independent of distance.

Assuming equal outgassing efficiencies for ^{14}N and ^{36}Ar on Mars, and taking into account losses of nitrogen from the upper atmosphere (McElroy et al., 1977), Pollack and Black concluded that the original $^{14}N/^{36}Ar$ ratio on Mars was 40 times that on the Earth, and from the absolute amounts of these two species in the Mars atmosphere they concluded that Mars had outgassed far less efficiently than the Earth. Taking terrestrial ratios for N_2/H_2O, they estimated that Mars had outgassed 6–160 m of water. This estimate suffers from the same drawbacks as the other chemical estimates so far discussed. It does not take into account losses by impact erosion and sputtering, and assumes that the various escape and exchange processes do not fractionate N_2 against H_2O. It also assumes the model of McElroy et al. (1977) for the evolution of nitrogen, which may be incorrect.

The The $^{15}N/^{14}N$ ratio in the Mars atmosphere is 1.62 times the terrestrial value (McElroy et al., 1977). This has been attributed to preferential loss of ^{14}N from the upper atmosphere as a result of a variety of photochemical mechanisms and ejection by oxygen ions picked up by the solar wind–induced magnetic field (McElroy et al., 1977; Fox and Dalgarno, 1983; Jakosky et al., 1994). The lighter isotope is preferentially lost because of diffusive separation of different molecular species above the homopause. From estimates of the loss rates of nitrogen, McElroy et al. (1977), attempted to reconstruct the history of nitrogen in the atmosphere, taking into account fixation into the ground and different mixing rates within the atmosphere. Their estimates for the original nitrogen inventory ranged from a few millibars to a few tens of millibars. Taking terrestrial values for the N_2/H_2O ratio, they derived the 6–160 m figure that was used by Pollack and Black (1979) for the outgassed water. Revised models of loss mechanisms suggest that McElroy et al. (1977) may have underestimated the loss rates. Fox (1993) estimates that the present observed enrichment of 1.62 could be achieved within 1.4 Gyr, so there is a problem as to why the nitrogen is not more enriched than is observed. Possible solutions are a thicker CO_2 atmosphere, particularly early in the planet's history, in order to decrease the N_2 mixing ratio in the upper atmosphere (Fox, 1993), progressive outgassing of N_2 (Jakosky et al., 1994), or exchange of N_2 with some surface reservoir, such as the regolith (Zent et al., 1994).

Estimates of H_2O from D/H Ratio

The deuterium to hydrogen ratio in the martian atmosphere is $8.1 \pm 0.3 \times 10^{-4}$, 5.2 times the ratio of 1.6×10^{-4} for terrestrial water (Owen et al., 1988; Bjoraker et al., 1989). This has intriguing possibilities for the history of water on Mars. It almost certainly implies that Mars had substantially more water in the past than at present. Enrichment occurs because of losses to space from the exosphere (Yung et al., 1988). Most of the

water in the atmosphere is in the lower two scale heights. In this part of the atmosphere photodissociation occurs and D and H are transported to the upper atmosphere as H_2 or HD. Above the homopause, at about 120 km, hydrogen and deuterium begin to separate because of their different masses, with the result that at the exobase, at about 240 km, a 10% enrichment of hydrogen is achieved. Hydrogen is preferentially lost with respect to deuterium, partly as a result of this diffusive separation and partly because of different effusion rates within the exosphere. Yung et al. (1988) examined the photochemistry and mechanics of hydrogen escape and estimated that the present escape rate of hydrogen is 1.6×10^8 atoms cm^{-2} s^{-1} and that of deuterium is 9.3×10^3 atoms cm^{-2} s^{-1}. The relative loss rates of D and H in proportion to their number densities in the bulk atmosphere is 0.32. The loss rates are very small because of the small amounts of water in the martian atmosphere. To accomplish the observed enrichment at these low escape rates, the reservoir undergoing fractionation must also be small. Yung et al. estimated that if the present rates were maintained for 4.5 Gyr, the observed enrichment would have been produced with an exchangeable water inventory of only 3.6 m spread over the whole planet, and of this they estimate that only 0.2 m remain.

As acknowledged by Yung et al., these low estimates appear to be in conflict with the geology. Two possibilities suggest themselves: Either the water in the atmosphere has been largely isolated from the rest of the water on the planet, or loss rates were higher in the past. I examined the first possibility (Carr, 1990). The atmosphere holds so little water that if it were isolated the observed enrichment would occur in about 10^5 years. If the atmosphere exchanged water with surface reservoirs with capacities of only 10–100 times that of the present atmosphere, the observed enrichment could occur in 10^6 to 10^7 years. I thus envisaged a punctuated water history with geologic events, such as floods and eruptions, episodically introducing juvenile water into the atmosphere, and this water was quickly frozen out at the poles. These events were followed by periods during which the atmosphere became enriched in deuterium because of only modest rates of exchange with surface reservoirs.

Jakosky (1990) alternatively suggested that the low inventories derived by Yung et al. were in error because they had assumed constant loss rates throughout the history of the planet. He pointed out that during periods of high obliquity sublimation at the poles is greatly enhanced and the water content of the atmosphere and loss rates from the upper atmosphere would rise correspondingly. Assuming that Mars spent 20% of its time at obliquities greater than 30°, and using models for the amount of water sublimed from the poles at high obliquity, he estimated that the time-averaged water content of the atmosphere is 20 times larger than the present water content. He concluded, therefore, that the inventories of Yung et al. should be multiplied by a factor of 20, thereby giving an original inventory of 72 m, with 4 m remaining.

My model involving little mixing between the atmosphere and surface now appears to be wrong in view of D/H measurements on water in SNC meteorites. Watson et al. (1994) measured the D/H ratio in biotite, apatite, and the hydrous magmatic mineral kaersutite, a Ti-rich amphibole, in several SNC meteorites (Figure 7-9). They found that all the samples had D/H ratios higher than terrestrial ratios and the highest ratios were comparable to the ratio in the martian atmosphere.* Ratios from a single mineral grain

*The D/H values are reported in terms of δD (per mil), where δD is defined as $[(^R\text{sample} - ^R\text{standard})/^R\text{standard}] \times 1000$. δD for the martian atmosphere is 4200.

showed considerable variability in D/H, and the ratio varied from mineral to mineral. A single kaersutite grain from Chassigny, for example, had δD values that ranged from 1253 to 1879 per mil. They ascribed the D/H enrichment to post-crystallization interaction of the samples with martian crustal fluids in an environment analogous to terrestrial hydrothermal systems. They ruled out a magmatic (mantle) cause for the high D/H, or assimilation of crustal materials by the magma because of the large intrasample and intramineral grain variability. Because the SNCs have ages ranging from 150 My to 1.3 Gyr, these results indicate that by 1.3 Gyr ago the D/H in surface waters (not just the atmosphere) had already evolved to close to its current value.

Interaction of martian magmas with meteoric water is consistent with oxygen isotope data. Karlsson et al. (1992) showed that the oxygen in water driven off from the SNC meteorites is isotopically distinct from the oxygen isotopes in the silicates. The silicates of all the SNC meteorites have oxygen isotope ratios distinctively different from those of the Earth and other meteorites. For a given $\delta^{18}O$ value, the SNC $\delta^{17}O$ values are consistently higher than terrestrial values by 0.3 per mil. However, $\delta^{17}O$ for the water in the SNC meteorites can range as high as 0.9 per mil over the terrestrial value. The high values indicate that the

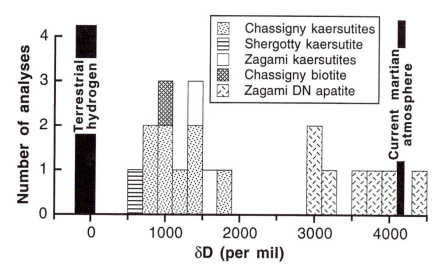

Figure 7-9. Histogram of δD values of individual hydrous phases in SNC meteorites. Bin size is 200 per mil. Uncertainties average ±68 per mil for kaersutites, ±158 per mil for the apatite, and ±40 per mil for biotite. The range of δD for terrestrial hydrogen and the current martian atmosphere are shown in black. The high and variable δD values may indicate that hydrous minerals containing unenriched mantle water were altered by enriched surface waters. (Reprinted with permission from Watson, et al., 1994. Copyright 1994 by the American Association for the Advancement of Science.)

water cannot come from the martian mantle, nor can it come from terrestrial contamination. Karlsson et al. concluded that the SNC magmas must have interacted with some crustal source of water that was isotopically distinct from the mantle. They proposed two possible alternatives as to why the crust and the mantle should be distinct. The first is that the crust and the mantle accreted from two isotopically distinct reservoirs that have remained separate since accretion. The second alternative is that after accretion the hydrosphere and mantle evolved independently. The hydrosphere was isotopically changed as a result of processes such as addition of cometary material or loss of oxygen from the upper atmosphere, while the mantle remained essentially the same. Mixing of the two reservoirs (mantle and crust) was trivial because of the lack of plate tectonics.

Thus both the D/H and oxygen isotopes suggest that SNC meteorites have reacted with surface water that is isotopically distinct from the mantle. Because the ages of the SNC meteorites range from 0.15 to 1.3 Gyr, these distinctions had to have been established by 1.3 Gyr ago. The D/H ratio gives an indication of how this surface reservoir might have evolved. For Rayleigh distillation.

$$\frac{I_{(0)}}{I_{(t)}} = \left(\frac{R_{(t)}}{R_{(0)}}\right)^{\frac{1}{(1-f)}}$$

where $I_{(0)}$ and $I_{(t)}$ are the total of hydrogen at times 0 and t, $R_{(0)}$ and $R_{(t)}$ are the D/H ratios at times 0 and t, and f is the fractionation factor, estimated by Yung et al. (1988) to be 0.32 for present-day Mars. In order to get an enrichment as high as 5.2, only a small fraction of the original inventory can be left (Figure 7-10). The present D/H enrichment implies that only 8% of the original water is left on the planet if the 0.32 fractionation factor has been maintained throughout martian history.

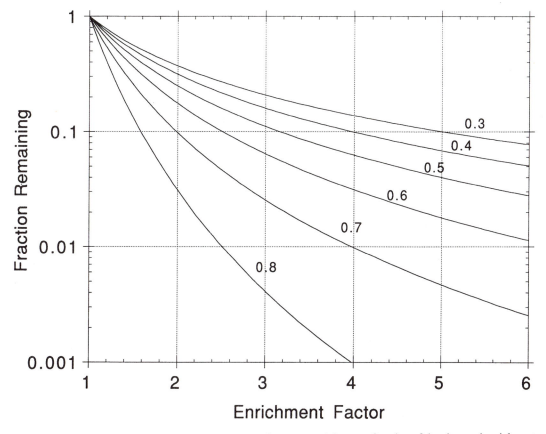

Figure 7-10. The fraction of the original inventory of water remaining as a function of the observed enrichment in D/H and different fractionation factors according to equation 7-1. For the observed 5.2 times enrichment in deuterium and the fractionation factor of 0.32 suggested by Yung et al. (1988), only 8% of the original hydrogen inventory remains.

There are two very different possible explanations for the D/H patterns seen in SNC meteorites. The first explanation (Jakosky and Jones, 1995) is that the atmosphere and the crustal water efficiently exchange with each other and that the present D/H ratio is the result of the preferential escape of hydrogen throughout the planet's history. This water has then reacted with SNC meteorites containing unenriched mantle water to cause the variable D/H values reported by Watson et al. (1994). The mechanism requires some means whereby atmospheric water can get into the groundwater system. They propose that this is effected in some way by hydrothermal activity, although precisely how this is achieved is left unstated. Introduction of significant amounts of atmospheric water into the ground at sites of hydrothermal activity under present climate conditions appears unlikely because the permafrost would be eliminated only over minute areas. Polar basal melting is another possibility, as discussed in Chapter 2. Alternatively, the climate has at times in the past been so different that the permafrost zone disappeared and significant seepage of surface waters into the ground could occur.

I find the Jakosky and Jones proposal unlikely on other grounds. As just duscissed, present loss mechanisms imply small total inventories of surface water. According to Jakosky (1990), even with the enhanced losses at high obliquity, the original inventory was 70 m and of this only 4 m remain. This appears quite inadequate to explain the abundant evidence for the action of water and ice discussed in earlier chapters. If indeed the total inventory of water at the end of heavy bombardment was significantly in excess of 70 m, as is implied by the geology, and the entire near-surface inventory is enriched in D/H as is implied by the SNC meteorites, then much more massive losses of water must have occurred in the past than are occurring during the current epoch, including the times of high obliquity.

A second possibility therefore is that massive losses of hydrogen early in the planet's history left the entire near-surface inventory of water enriched in D/H. This is essentially the explanation given by Owen et al. (1988) when they first discovered the D/H enrichment in the martian atmosphere. They suggested that Mars had a significantly warmer climate early in its history, that the atmosphere contained much more water than at present, and that the hydrogen loss rates were correspondingly higher. Major losses of hydrogen during this era left the remaining water enriched in deuterium. The inter-sample and intra sample variabilities in D/H are straightforwardly explained if the primary minerals containing unenriched water from the mantle are partly altered by enriched surface waters. According to this explanation, the identity between the atmospheric value and the maximum value in the SNCs results because present day losses of water are small compared to the size of the reservoir being fractionated. The reservoir exchanging with the atmosphere and currently being fractionated could include ice within the permafrost and the polar layered terrains.

Donahue (1995) pointed out that the near-identity of the D/H in the Zagami meteorite with that in the present-day atmosphere provides a means of placing limits on the amount of water on the planet today and in the past. The estimated precision of the D/H measurements in the meteorities is 4 percent. If Zagami is 1.3 Gyr old (McSween, 1985), then the atmosphere must be exchanging with a reservoir containing the equivalent of at least 25 m spread over the whole planet otherwise the D/H in Zagami would be detectably less than in the atmosphere. The original exchangeable surface reservoir had to be at least 280 m to provide the 5.2 enrichment. These are lower limits if Zagami is 1.3 Gyr old and the fractionation factor of 0.32 is valid for the history of the planet. Clearly, if the Zagami D/H values are actually within 1 percent of the present day atmosphere, then the equivalent of 1.1 km could have been present early, and 100 m left today. Donahue (1995) emphasized that in order for the planet to lose large amounts of water, hydrogen loss rates had to be much higher in the past. If most of the original 255 m were lost in the first 1 Gyr, hydrogen loss rates had to be close to 10^{11} atoms cm^{-2} sec^{-1} as compared with loss rates of 2 x 10^{-8} cm^{-2} sec^{-1}. He suggested that such high loss rates could have been caused by warmer and wetter climates and by heating of the upper atmosphere by the enhanced UV output of the young Sun. The losses must have been accompanied by enhanced losses of oxygen which could also been stimulated by the enhanced UV solar flux.

If geologic estimates of the amounts of water

that were involved in channel formation during middle Mars history are true, then even larger amounts of hydrogen may have been lost from the planet than estimated by Donahue (1995). For example, if the minimum inventory of 400 m estimated from water erosion is correct (see below), and the fractionation factor of 0.32 is correct, and all of it is enriched, then the equivalent of at least 5 km of water was lost early in Mars' history. We have already seen that Mars is likely to have lost much of its original water during core formation. Water incorporated into the interior would have reacted with metallic iron releasing massive amounts of hydrogen to drive hydrodynamic escape. The amounts of water lost may be equivalent of several tens of kilometers or more (Dreibus and Wänke, 1987; Pepin, 1991). However, the fractionation factor for D/H during hydrodynamic escape is estimated by Zahle et al. (1990) to have been in the range of 0.8–0.9. With these fractionation factors enormous amounts of water would need to be lost to leave 0.4 km enriched by a factor of 5.2. But as the hydrogen loss rate declined, so would the fractionation factor, as diffusive separation between D and H in the upper atmosphere became more efficient. Ultimately, the fractionation factor would have declined to its present value of 0.32. Unfortunately, the scenario just outlined has not been modeled, and it is not clear whether the combination of hydrodynamic escape very early in Mars' history and warm climates during the era of heavy bombardment could have resulted in a 5.2 enrichment of the entire, and rather substantial, near-surface inventory of water that is required to explain the geology.

Geologic Estimates of the Surface Inventory of Water

Geologic estimates for the amount of water at the surface have tended to be larger than other estimates. The estimates are based mainly on three considerations: the holding capacity of the megaregolith for water, the amount of water required to cut the erosion features observed, and the amount of water required to produce the large bodies of water postulated to have pooled in parts of the low-lying northern plains. As discussed in the previous chapter, the amounts of water that reside in readily identifiable reservoirs are small.

As can be seen from Table 6-1, the maximum likely in the atmosphere, regolith, and various polar reservoirs combined is probably no more than 40 m spread over the whole planet and could be considerably less. If Mars has or had a large inventory of water, as appears likely, then it is in the cryosphere or the underlying groundwater system. A reasonable upper limit on what these two reservoirs might hold is provided by the 50% surface porosity megaregolith of Clifford (1993), which suggests a holding capacity of 1400 m spread over the whole planet.

Estimates can be made of the amounts of water that were at one time present in the megaregolith from the amount of water erosion (Carr, 1986). The large outflow channels around the Chryse basin have removed approximately 4×10^6 km^3 of material to form the outflow channels, chaotic terrain, and smaller canyons (Table 3-2). If we assume that all the water that flowed through these channels carried the maximum reasonable sediment load (40% by volume, as discussed in Chapter 3), then the minimum amount of water that flowed through these channels is 6×10^6 km^3, or the equivalent of 40 m spread over the whole planet. To determine what this implies about the total surface inventory, we must make some judgment as to whether outflow channels started at specific locations because water was especially abundant in that part of the crust, or whether they started because the conditions necessary for rapid release were achieved. In view of the almost universal evidence for water erosion in the old terrains (see Figure 4-2) and the likely interconnectivity of the global water system, it is more likely that water was widely distributed, and the preferential location of outflow channels around Chryse, and in Elysium, Hellas, and Memnonia, is because the special conditions required for rapid release of water were met in these places. Estimating that the area drained by Chryse channels constitutes roughly one tenth of the planet's surface, the Chryse water volumes can be extrapolated to the entire planet to derive a global total of 400 m of water.

This estimate is very crude and little more than a guess. On the one hand, the estimate is too high if there was some recycling mechanism and the same water passed through the channels more than once, or if subsurface migration of water resulted in unusual concentrations of groundwater in the source regions of the channels. On the other hand, the estimate is too low if all the water did not carry its maximum sediment load, or if significant amount of water were left in cryosphere and groundwater system after the floods, conditions that are both likely. Although the estimate is crude, a near-surface inventory of less than a few hundred meters of water appears unlikely in view of the amount of water erosion that has been achieved.

The former presence of ocean-scale bodies of water postulated by Baker et al. (1991) obviously implies larger inventories. The volumes of water suggested by their map is closer to the higher values listed in Table 6-3 than the lower values. These volumes represent only the water on the surface, not that remaining in the cryosphere or the groundwater system, so that total surface inventories of well over 1 km are probably required.

Another geologic estimate of the amount of water at the surface is based on the volume of volcanic materials erupted. The estimates of volumes of erupted volcanic rocks (Figure 7-8) is based on the areas of different-aged volcanic units on global geologic maps (Scott and Tanaka, 1986; Greeley and Guest, 1987) and on thicknesses estimated from partial burial of craters. Assuming a terrestrial ratio of erupted to igneous rocks of 10:1, Greeley and Schneid (1991) estimated that approximately 650×10^6 km^3 of magma have been brought close to the surface since the mid-Noachian. Greeley (1987) somewhat arbitrarily assumed that martian basalts would have the same water contents as terrestrial basalts (1%) If so, Mars has outgassed roughly 150 m of water since the mid-Noachian. This is probably an upper limit on the water outgassed to the atmosphere since that time, because it is unlikely that all the water brought up in magmas outgassed, leaving nothing behind in the rocks. Furthermore, the martian mantle may be drier than that of the Earths (Dreibus and Wänke, 1987) and the volcanic magmas may have brought correspondingly less water to the surface.

Summary

The foregoing discussion presents a number of seemingly conflicting estimates of the amount of water acquired by Mars (Figure 7-11). Nevertheless, a reasonably consistent picture appears to be emerging. Mars accumulated from more volatile-rich materials than the Earth, as expected from its greater distance from the Sun and as indicated by the chemistry of the SNC meteorites. It would initially have incorporated proportionately more water into its interior than the Earth, because of its more volatile-rich source materials and because its smaller size would have resulted in a smaller fraction of the accreting materials being devolatilized on impact. Toward the end of accretion there may also have been significant amounts of water present at the surface, either in a primitive atmosphere or condensed on or below the surface. At or near the end of accretion, the planet differentiated, metallic iron reacted with the water within the interior, and a massive hydrogen loss resulted. At about the same time dust and gas were cleared from the solar system and the planet became exposed to high fluxes of EUV from the early Sun. The result was large losses of hydrogen from the upper atmosphere, possibly partial enrichment of the surface waters in deuterium, and large losses and fractionation of other volatiles as a consequence of hydrodynamic escape.

A major uncertainty at this stage is the completeness with which water was lost from the planet. Dreibus and Wänke (1987) argued that the interior was left with only 36 ppm, that surface water was eliminated, and that the subsequent water erosion was effected by partial outgassing of only a few tens of meters of water from the interior. This seems improbable, for a few tens of meters of water is almost certainly too small an amount to explain the large outflow channels and other indicators of water and ice. There are several possibilities whereby we can reconcile the apparent retention of large amounts of water with the chemical evidence for mantle equilibration with the core. The first is that impact devolatilization left large amounts of water near the surface, either in the atmosphere or condensed in or on the surface and that, although a large fraction of this may have subsequently been lost, a few hundred meters could have been retained as appears required by the geology. A second possibility is that not all the water in the interior was consumed by reaction with iron. A small fraction may have been left to be subsequently

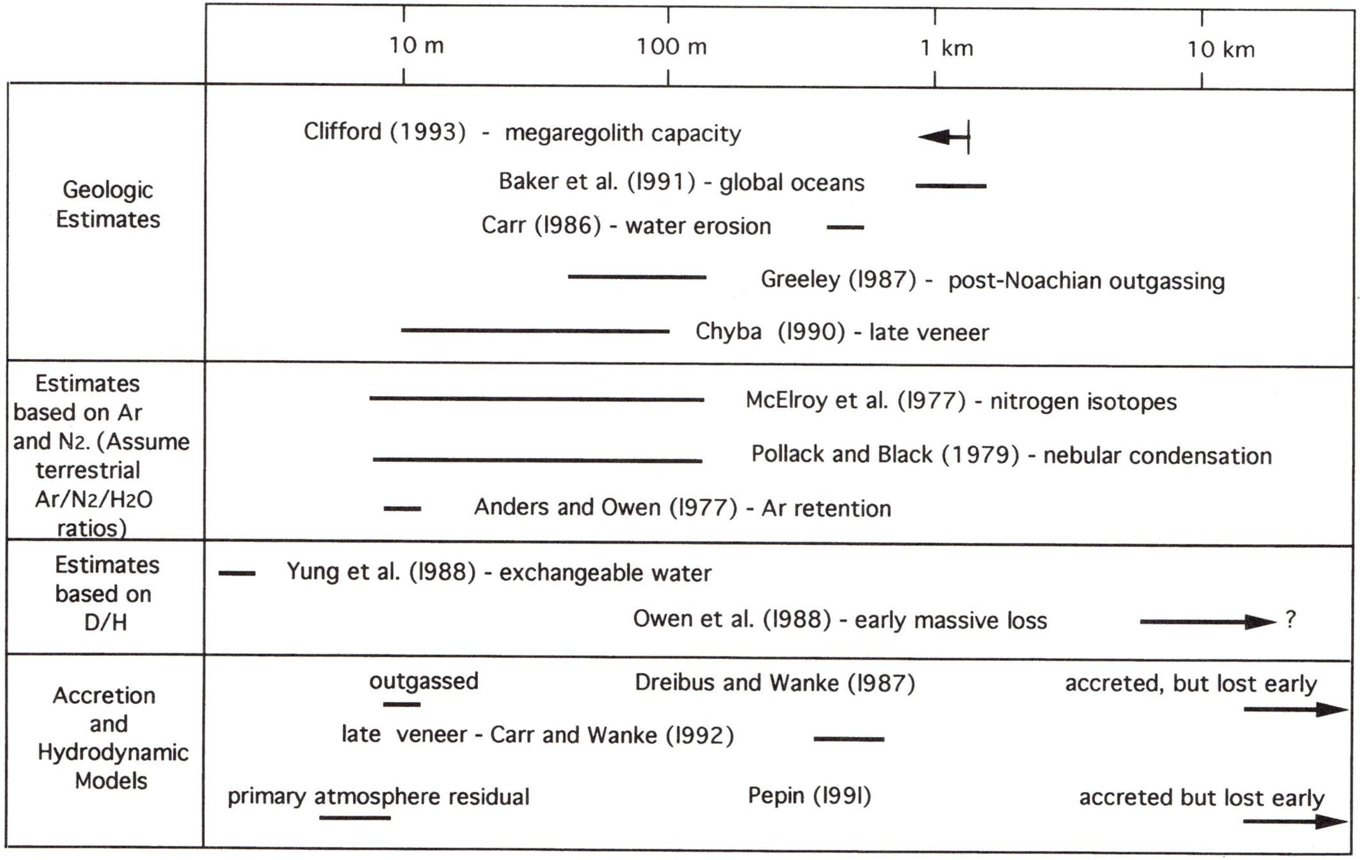

Geologic Estimates

10 m	100 m	1 km	10 km

Clifford (1993) - megaregolith capacity

Baker et al. (1991) - global oceans

Carr (1986) - water erosion

Greeley (1987) - post-Noachian outgassing

Chyba (1990) - late veneer

Estimates based on Ar and N2. (Assume terrestrial Ar/N2/H2O ratios)

McElroy et al. (1977) - nitrogen isotopes

Pollack and Black (1979) - nebular condensation

Anders and Owen (1977) - Ar retention

Estimates based on D/H

Yung et al. (1988) - exchangeable water

Owen et al. (1988) - early massive loss   ?

Accretion and Hydrodynamic Models

accreted, but lost early

Dreibus and Wanke (1987)

late veneer - Carr and Wanke (1992)

Pepin (1991)

outgassed

primary atmosphere residual

accreted but lost early

Figure 7-11. Estimates of different amounts of water on Mars as discussed in the text. Amounts are expressed in terms of the thickness of a layer spread evenly over the whole planet.

outgassed. Most of the outgassing probably took place early, when heat flows were much higher than at present. The amount of water originally incorporated into the planet is likely to have been very high, possibly the equivalent of over 100 km spread over the whole planet, so only a small fraction needs to have been left unconsumed by Fe to be subsequently outgassed and form the waterworn features. A third possibility is that Mars, like the Earth, acquired a volatile rich veneer early during heavy bombardment, but this was not mixed deep into the planet because of the lack of plate tectonics. As a consequence, no siderophile anomaly was left in the mantle.

Survival of the hundreds of meters of water thought required to produce the water erosion features can, therefore, be reconciled in different ways with what we know about accretion, core formation, and heavy bombardment. Carr and Wänke (1992), in an attempt to reconcile the seemingly water-rich surface with the apparent equilibration of the mantle with the core, suggested that, after accretion, Mars may have been left with a dry interior but a water-rich surface, and the two have been largely unmixed ever since. This suggestion appears supported by the hydrogen and oxygen isotopics and the low water content of the SNC meteorites, although these all have other possible explanations.

The early massive losses of hydrogen could have left the surface waters enriched in deuterium. Although fractionation during the main hydrodynamic phase is likely to have been inefficient, significant enrichment could have been achieved during the waning phases of hydrodynamic escape as the fractionation efficiency increased with the declining loss rates. Further fractionation may have occurred as a result of losses from the thick, water rich atmosphere postulated for the period of heavy bombardment on the basis of high erosion rates and high rates of valley formation. Loss rates may have been further amplified by the enhanced UV output of the early Sun.

Once the era of hydrodynamic escape and massive outgassing of the interior was over, which Pepin (1991) estimated to be about 4.2 Gyr ago, losses and additions to the surface inventory would have been modest. Continued outgassing of the interior could have added an additonal 100 m, thereby diluting any deuterium

enrichment of the surface water. This may be an upper limit on the amount of water outgassed by post-Noachian volcanism in view of the low water content of the SNC meteorites as compared with terrestrial basalts. Post heavy bombardment losses of water from the upper atmosphere probably amount to no more than a few tens of meters, thereby adding slightly to the deuterium enrichment.

Water in SNC meteorites contains oxygen that is isotopically distinct from the oxygen in the silicates. The oxygen reservoir at the surface and the oxygen reservoir of the mantle appear to have evolved independently with the minimum of mixing, again possibly because of the lack of plate tectonics. The variable D/H ratios in water extracted from SNC meteorites can similarly be attributed to different proportions of an unenriched mantle and an enriched surface component, although other explanations are possible. The near-surface waters must have acquired their deuterium enrichment and unique oxygen isotopic compositions by 1.3 Gyr ago. Some of the SNC meteorites appear to have undergone hydrous alteration at temperatures of several hundred degrees Kelvin (Watson et al., 1994), which implies hydrothermal activity; other alterations seemingly took place at much lower temperatures (Gooding, 1992). Thus, water affected by processes at the top of the atmosphere was able to penetrate deep enough into the crust to participate in hydrothermal activity. This is easily accomplished if the isotopic signatures were imposed on surface waters very early, because water (and carbonates) would have been folded deep within the crust as a consequence of heavy bombardment and/or water would have been able to percolate from the surface into the crust because of warm surface temperatures. If the isotopic signature was acquired relatively late (but still prior to 1.3 Gyr ago), then the presence of isotopically fractionated water deep within the crust is more difficult to explain because surface conditions had to allow water in the atmosphere to penetrate below what is now a deeply frozen surface. Mechanisms such as basal polar melting (Clifford, 1993), episodic climate change (Baker et al., 1991), or local seepage at the site of hydrothermal activity would be needed.

We can conclude from this section that the absolute amounts of noble gases in the atmosphere are unlikely to provide reliable estimates of the amount of water at the surface because the noble gases, being mostly in the atmosphere, have been subject to loss processes not experienced by most of the water, it having been mostly in or on the surface. [40]Ar may provide a reliable means of estimating the efficiency of outgassing of the interior after core formation but the implications for the water story are unclear because we do not know how much water remained in the interior by the time global differentiation was over. Estimates of water inventories by reconstructing the evolution of the atmosphere from nitrogen isotopes is subject to comparable, although different, difficulties.

8 IMPLICATIONS FOR LIFE

The increasingly compelling evidence that Mars is water rich, and that it may have experienced major changes in its climate, coupled with recent improvements in our understanding of the early evolution of life on Earth and the range of conditions under which it can survive, has led to a revival of interest in the possibility that life may have started on Mars. Liquid water is almost universally accepted as a requirement for life, and we have almost unequivocal evidence that liquid water was at the surface on many occasions. High erosion rates, high rates of dissection to form valley networks, and high rates of weathering all suggest, although with considerable uncertainty, that conditions, at least on early Mars, were warm in addition to being wet. Episodic formation of outflow channels throughout the planet's history indicates the presence of abundant groundwater.

Volcanism has also continued throughout the planet's history. The volcanics must have commonly interacted with groundwater and ground ice to cause hydrothermal activity, and this appears supported by the mineralogy and isotopics of SNC meteorites. Large floods, an extensive groundwater system, and the presence of layered sediments imply that lakes, and possibly larger bodies of water, formed at various times. Thus, not only may early conditions on Mars have been conducive to the origin of life, but if life did start there were numerous niches in which living organisms might have survived. The likelihood of persistent hydrothermal activity on Mars is particularly intriguing because recent studies in molecular phylogeny indicate that the most primitive, least evolved organisms on Earth live in high-temperature, hydrothermal regimes, utilizing geothermal energy sources such as sulfur and molecular hydrogen (Pace, 1991, Woese, 1987).

The main objective of the Viking mission was to look for life. Each lander had a threefold biology experiment and a gas chromatograph/mass spectrometer (GCMS). The biology experiments were designed to detect microbial growth in different ways. The pyrolytic release experiment detected incorporation of CO and CO_2 into organic compounds. The gas exchange experiment measured uptake and release of various gases. The labeled release experiment detected decomposition of simple organic compounds. While each of the experiments yielded results that were not anticipated for a nonbiologic setting, the consensus among the experimenters was that the results had plausible, nonbiologic explanations (see, for example, Klein, 1978, 1979; Horowitz, 1986, and for an opposite view, Levin and Straat, 1981; Levin, 1988). The lack of detection of organic molecules by the GCMS, despite detection limits of the order of parts per billion (Biemann et al., 1977), and the oxidizing nature of the soil support the nonbiologic interpretation.

In retrospect, even if life has survived today on Mars, the probability of detecting it at the two Viking landing sites is probably small. The Sun's ultraviolet radiation, longward of 2000 Å, passes almost unattenuated through the martian atmosphere (Figure 8-1). We know that such radiation is very harmful to terrestrial organisms (see, for example, Jagger, 1985). Presumably, any martian life would be based on complex organic chemistry and be similarly vulnerable. Second, we know that the soils exposed to the atmosphere are oxidizing, which, in combination with the ultraviolet radiation, leads to destruction of organic compounds. Yet the pyrolytic release and labeled release experiments were designed to detect heterotrophs, that is, organisms that metabolize organics. Third, liquid water is not available anywhere on the surface to enable metabolism. The expectation was that living forms might be dormant, awaiting the occasional event, such as a flood or melting of polar ice at high obliquity, that would make liquid water available so that they could revive and multiply. However, dor-

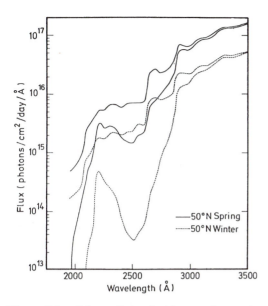

Figure 8-1. Solar radiation incident on the martian upper atmosphere (upper lines) and surface at 50°N. The adsorption at 2500Å in the winter hemisphere is due to the presence of small amounts of ozone. On Earth the flux drops precipitously below 2900Å because of ozone. (From Kuhn and Atreya, 1979. Reproduced with permission from Academic Press.)

mancy at the surface must include some efficient repair mechanism or protection against UV-induced damage. If life did start on Mars and did survive to the present, it is unlikely to be at the surface. Far more probable is survival in protected places, such as hydrothermal systems or aquifers, wherever liquid water and the appropriate nutrients are available.

In this chapter, we review what little is known about how life started on Earth and the conditions under which it might have started. We also examine some of the limits to survival of terrestrial life. We then examine some possible martian environments where the postulated conditions for starting life might have been met.

Origin and Early History of Life on Earth

The oldest unequivocal life forms on Earth are stromatolites from 3.5-Gyr-old sediments in South Africa and Western Australia (Schopf and Walter, 1983; Schopf et al., 1983). We saw in the previous chapter that as a result of core formation and burial of accretional energy, the Earth probably had a massive (100 bar) steam atmosphere at the end of accretion. Surface temperatures may

have been in excess of 1500 K, the surface had probably melted, and the mantle was actively convecting. As the accretion rate slowed, the surface cooled and the mantle started to crystallize. At this stage the Earth probably acquired a thin, unstable lithosphere as a result of freezing of the molten silicate liquid at the surface and upward flotation of crystals from below (Ernst, 1983). Meanwhile, the steam atmosphere collapsed to form the oceans. As the planet cooled further, the lithosphere would have thickened, but it likely was repeatedly disrupted and reingested into the mantle as a result of the vigorous mantle convection. The outermost crust would not have become consolidated at its base until basal temperatures had fallen to about 1000 K, which is estimated to have taken about 0.3 Gyr (Hargraves, 1976). As the mantle cooled, injection of sialic residual melts, rich in incompatible elements, into and onto the lithosphere would have ultimately led to a lithosphere covered with a veneer of sialic crustal material. Large impacts would have repeatedly disrupted the crust and evaporated large volumes of oceans. Whether life could or did start under these conditions is unclear, but repeated large impacts would likely have frustrated any early attempts at life (Sleep et al., 1989).

The 3.8-Gyr-old metasediments of Isua, Greenland provide the first concrete record of conditions on the early Earth. The Isua rocks consist of various kinds of amphibolites and of metasediments such as metacherts, carbonate-bearing mica schists, and metaconglomerates with volcanic clasts (Ernst, 1983). They indicate that by 3.8 Gyr ago, there were extensive bodies of water and protocontinents, that surface temperatures had fallen below 100°C, and that the atmosphere was CO_2 rich and anoxic. Modeling of early climates (Kasting, 1987; Kasting and Toon, 1989) indicates that at least several tenths of a bar of CO_2 would be needed to keep surface temperatures at the present level, but as much as several bars may have actually been present. These amounts of CO_2 are required to maintain warm surface temperatures because of the reduced solar constant, as discussed in Chapter 6. The total surface inventory of CO_2 is estimated to be 3×10^{20} kg, or about 60 bars, and almost all of it is now sequestered in limestones (Holland, 1978). Although no living forms have been found in the Isua sediments, Cloud (1968, 1983) has argued that banded iron formations found within the sequence formed by precipitation of iron as a result of photosynthetically produced oxygen. In addition, biologic reactions commonly fractionate the isotopes of carbon (Schidlowski, 1987) such that most organic carbon has $\delta^{13}C$ values of -10 to -40 per mil. Kerogen in the Isua sediments is similarly depleted, but inorganic causes are possible (Pace, 1991).

The oldest identified living forms are in 3.3 to 3.5-Gyr-old sediments from South Africa and Western Australia (Figure 8-2). These contain structures that have been identified as microfossils and stromatolites (Walter, 1983; Schopf and Walter, 1983). Stromatolites are banded formations readily visible to the unaided eye. They form when layers of microbial organisms at the bottom of shallow lakes or tidal lagoons are covered with sediments. Because the sediments obscure the sunlight, the photosynthesizing organisms move upward to survive and the rest of the community follows, leaving organic detritus behind. This material combines with the inorganic sediments to form microbial mats, which on lithification form stromatolites. They form today only in extreme environments, such as hypersaline ponds and ice-covered Antarctic lakes, where bottom grazers that would feed on the mats cannot survive. They are important for exploration, both of early Earth and Mars, because although they are microbial in origin they are easily recognizable macrostructures.

Thus life on Earth arose remarkably quickly. Almost immediately after the end of the era of large and possibly sterilizing impacts, there are suggestions of metabolism. By 3.5 Gyr ago, within 300 million years of the end of heavy bombardment, stromatolite colonies were flourishing in different parts of the world. By this time, life appears to have been quite diverse and complex. Analogies with ancient and present-day stromatolite communities suggest that these early ecosystems included both heterotrophic organisms (those that use organic matter as a source of cellular carbon) and autotrophic organisms (those that use CO_2 as a carbon source), and that some were capable of using sunlight for energy (Chang, 1982). All were probably anaerobic.

Most speculation on the origin of life has focused on the Oparin-Haldane model (Oparin, 1957; Haldane, 1929) in which living systems

rose naturally through chemical evolution of organic matter. For the process to start, simple organic materials must be available. The simple organic starting compounds could have been derived from various sources. Experiments by Miller (1953, 1955, 1957) showed that a wide range of organic compounds are synthesized by passage of an electric discharge through various reducing atmospheres (CO_2-N_2-H_2O, CO-N_2-H_2O, etc.). Similar arrays of compounds (Table 8-1) were subsequently synthesized by using ultraviolet light and by passing the putative atmospheric gases over different mineral catalysts (see Chang, 1982; McKay, 1991 and references therein). In the less reducing or neutral atmospheres, yields drop significantly. Studies of the evolution of planetary atmospheres and the photochemistry of ammonia and methane suggest that the early atmospheres of Venus, Earth, and Mars would not have been reducing, but rather dominated by CO_2, N_2, and H_2O (Owen et al., 1979; Pollack and Yung, 1980; Walker, 1985). NH_3 and CH_4, whose presence in the Miller experiments results in rich organic yields, would have been rapidly destroyed by photolysis. However, during the final stages of accretion, the atmosphere may have become more reducing as a result of a higher influx of chemically reduced cometary and asteroidal material (Kasting, 1987).

Even if organic compounds were not synthesized in quantities on the early Earth because of the neutral atmosphere, they should have been readily available from exogenous sources such as comets and asteroids (McKay, 1991; Chyba and Sagan, 1992). Cometary material is rich in organics (Kissel and Krueger, 1987) and so are carbonaceous chondrites (Cronin, et al., 1988). One uncertainty is the extent to which these organics would survive entry into the Earth's atmosphere. We have numerous examples in our meteorite collections of organics preserved in asteroids, so we know that extraterrestrial organics are delivered to the surface of the Earth. Chyba and Sagan argue that significant amounts of the organics in comets would also be preserved, because of disintegration of these poorly cohesive objects in the atmosphere. Preservation would be favored by the thicker atmospheres expected on the early Earth. Organics would also be synthesized in the atmosphere by shock heating of the reduced gases that result from disintegration of the

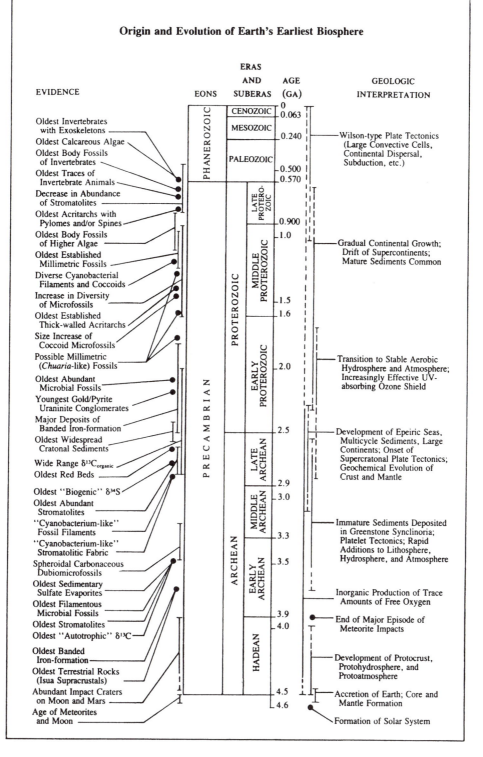

Figure 8-2. A synopsis of the evidence for the origin and evolution of life on Earth and some of the accompanying geologic events. (From Schopf et al., 1983. Reproduced with permission from W. Schopf, UCLA.)

Table 8-1. Results of experiments on abiotic synthesis (From Chang, 1982).

Gases	Electrical discharges	Ultraviolet photochemistry	Fischer-Tropsch Process
CO_2, N_2, H_2O	HNO_3, etc. (?)	Formic acid, formaldehyde	
CO_2, CO, N_2, H_2	HCN, amino acids, etc.(?)		
CO, N_2, H_2O		Aldehydes, alchohols, acids, ketones	
CO, NH_3, H_2	HCN, amino acids, etc.		Hydrocarbons, aldehydes, ketones, nitriles amino acids, purines, etc.
CH_4, N_2(NH_3), H_2, H_2O	HCN, hydrocarbons, aldehydes, ketones, amino acids, carboxylic acids, etc.	Aldehydes, alchols. ketones, hydrocarbons (HCN)[a] , amino acids[a]	

[a] Formed when N_2 is replaced by NH_3.

comets. Other organics would be delivered to the upper atmosphere in interplanetary dust particles with minimal heating. Thus, the early Earth was probably amply supplied with simple organic compounds, as Mars would also have been.

We do not know, of course, how life evolved from this mix of simple organics. Generally what is envisaged is a sequence resembling the following (Chang, 1982). First came simple monomers such as amino acids and nucleotides. From the monomers were formed oligomers of higher molecular mass and polymers, such as oligonucleotides and peptides, with rudimentary catalytic properties. Next the oligomers and polymers, various organic molecules, and inorganic complexes may have accumulated into droplets enclosed by primitive, lipid membranes. Finally, isolation within the membrane-bound environment allowed the different components to evolve to achieve the functional complexities of metabolism and replication. This is all very vague, but the fact that it happened so quickly on Earth lends credence to the supposition that it could have happened on Mars if the physical and chemical circumstances were similar.

Even if life did not start on Mars, exploration of plausible fossil environments on Mars where it might have started is of considerable interest because of evidence that some of the steps in the prebiotic chemical path just described may be preserved. On Earth, the chances of finding direct evidence of prebiotic chemistry is almost nil because the planet has been so warm, wet, and active all of its history and because the effects of life itself are so pervasive. The evidence would be much more readily preserved on Mars as the planet has been cold for most of its history and much less active.

Molecular Phylogeny and the Origin of Life

In the last decade important clues about the possible physical setting in which life started have been obtained from studies of the sequencing of nucleotides in RNA (Woese, 1987; 1990; Pace, 1991). The RNA molecule is a long chain of nucleotides. Each nucleotide consists of a ribose sugar, a phosphate group, and one of four bases (adenine, uracil, guanine, and cytosine). Individual sections of RNA are identifiable by their unique sequence of nucleotides. The sequences are inherited, with slight modifications, in the process of replication. Thus, the similarity in the sequences on specific types of RNA in different organisms is a measure of how closely related the organisms are in the evolutionary tree. Woese and his coworkers chose to examine the sequencing on ribosomal RNA in order to determine quantitatively the evolutionary relatedness of different organisms. Ribosomes are small structures within all living cells that are thought to control the translation of genetic information into the production of proteins. They are fundamental to the basic metabolism of the cell. Ribosomal RNA was chosen for the evolutionary studies partly because of its abundance (a typical bacterial cell contains 10,000-20,000 molecules), partly because it appears to have retained its same function over large evolutionary distances, and partly because the sequencing appears to have been modified only slowly so that common, distant, ancestral sequences are not completely obliterated. The particular molecule they examined, 16S ribosomal RNA, is about 1540 nucleotides long.

Phylogenetic trees based on rRNA show that all organisms are related and that they must have had a common ancestor (Figure 8-3). Three primary lineages are designated as Archea, Eucarya and Bacteria (Woese, 1990), in contrast to the previous twofold classification of Eukaryotes and Prokaryotes. The root of the phylogenetic tree, although not firmly established, is thought to be attached to the Archea branch.

From the perspective of Mars exploration, what is of interest here are the characteristics of the most primitive organisms that are recognized, because they may provide clues as to the conditions under which life started on Earth. Sequence comparisons show that the different evolutionary lineages have not evolved at the same rates. Because their last common ancestor, sRNAs of the Archea, have accumulated far fewer mutations than the Eucarya or Bacteria, they are much more similar to the common ancestor. Many of the Archea have unusual properties. Methanogenic Archea live only in oxygen-free environments and generate methane by reduction of CO_2; they are thus ideally suited to the atmosphere thought to have existed on the primitive Earth (and Mars). Methanogens are widely distributed in

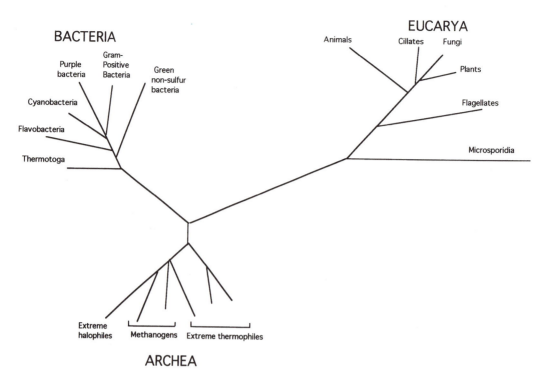

Figure 8-3. Universal phylogentic tree as determined by ribosomal RNA sequence comparisons. The lengths of the branches are a measure of the evolutionary distances as recorded in the difference in the RNA sequences. The Archea branches are shorter and closer to the common ancestor at the junction of the three main stems and thus are thought to most closely resemble the ancestor. (Adapted from Woese, 1987.)

anaerobic situations, such as bogs, animal intestines, and sewage plants. They are also found in hot springs. Other Archea are halophiles, able to live in very salt-rich environments; others are thermoacidophiles that live in acid waters in hot springs.

The most primitive life forms on Earth appear to be anaerobic hypothermophiles that grow at temperatures of 90°C or higher, utilizing geochemical energy sources such as sulfur and hydrogen, and fixing CO_2 (Pace, 1991; Achenback-Richter et al., 1987). In many forms, growth ceases when temperatures drop below about 80°C. As a consequence, Pace (1991) suggested that the universal ancestor is likely to have been a chemosynthetic thermophile that lived in hot springs, rather than a heterotroph living at low temperatures in a more benign environment, such as a tidal pool. Pace also pointed out the difficulty of synthesizing RNA in aqueous solutions, as required by the chemical evolution model outlined earlier because of the susceptibility of RNA to hydrolysis. He suggested that surface chem-

istry on clays (Cairns-Smith, 1982), pyrite, or basaltic glasses (Wächterhäuser, 1988), with low water activities, may have been more important in the creation of replicative mechanisms than polymerization in water-borne droplets. The relevance of these recent findings for molecular phylogeny on Mars is clear. Given that early Mars was water rich and highly volcanic, hot springs must have been common. Yet this is the environment in which the most primitive life forms on Earth live and the one in which the molecular evidence suggests terrestrial life most likely originated.

Life in Extreme Environments

The adaptability of terrestrial life to extreme environments is also of interest with respect to Mars because the extremes bracket the conditions under which terrestrial-like life could survive. If life did start on Mars, it may have done so under conditions very different from those that prevailed for most of the planet's subsequent history. To survive, living forms would have had to adapt to the changing conditions. The ability of life to survive and adapt to seemingly hostile conditions is also of interest from two additional perspectives: natural contamination from the Earth and human-induced contamination. SNC meteorites almost certainly come from Mars. The fact that we have identified so many meteorites that come from Mars within the last 100 years suggests that infall of martian material on Earth is not unusual, and that the Earth has been sprinkled with martian rocks throughout much of its history. The reverse process is more difficult because of the Earth's larger gravitational field. Nevertheless, over the last few billion years terrestrial rocks must have frequently fallen on Mars, and during heavy bombardment such showers must have been common. Many of the SNC meteorites are only mildly shocked, and survival of organisms through ejection from the Earth appears likely. But could the organisms survive the long transit times to Mars, which are likely to be on the order of 10^5 to 10^6 years? Survivability is also a concern with respect to biologic contamination of the planet, because in order to do more definitive biologic experiments on Mars, we need to be careful that experiments are not jeopardized by terrestrial contaminants (National Research Council, 1992). Introduction of some contaminants is impossible to avoid, so we wish to better understand how such an introduced biota might survive under martian conditions.

Terrestrial life can survive in a remarkably broad range of conditions. We have already seen that it can survive in extremely hot temperatures. Extremely thermostable enzymes have, for example, recently been extracted from hyperthermophiles from oceanic vents. An amylase from the Archea *Pyrococcus furiosus*—a heterotroph capable of growing at temperatures up to 103°C—is active at 140°C (Koch et al., 1990). Other enzymes from the same organism show optimal activity in the 105–110°C temperature range. If liquid water is present at kilometer depths on Mars, as appears likely, microbial life may encounter environmental conditions similar to those experienced by barophilic bacteria found in the deep sea (Yayonos, 1986). Deep waters on Mars are expected to be saline. Extremely halophilic Archea have been cultured from the interior of salt crystals approximately 200 Myr old from salt deposits in New Mexico (National Academy Sciences, 1992), demonstrating both the extreme survivability and the salt tolerance of these organisms.

Cryptoendoliths from Antarctica demonstrate another form of tolerance and adaptability. These organisms have adapted to extremes of low temperature, high winds, and lack of water by forming communities within sandstones (Friedmann, 1982). Inside the rocks, light penetration is sufficient for photosynthesis, and the enhanced relative humidity occasionally provides adequate liquid water for growth. The communities are complex, consisting of lichen, green algae, cyanobacteria, yeast, and filamentous fungi. Their rocky host provides protection against the very harsh external conditions, which include high winds, temperatures that never rise above freezing, low moisture, and high UV fluxes. The communities can carry on photosynthetic metabolism at temperatures as low as −8°C, and some lipids can stay fluid at temperatures as low as −20°C.

Perhaps the condition on Mars that is most adverse for survival is the UV radiation. Spores are known to survive for thousands of years and be resistant to freezing, desiccation, vacuum, and disinfectants. But experiments have shown that when spores are not shielded from solar radiation, the survival rate drops by two to four orders of magnitude (National Research Council, 1992). The results suggest that survival in the vacuum and cold conditions of space is not a problem, but in the absence of UV shielding there is little chance for surviving for very long. The same is true of the surface of Mars. This appears true despite adaption of a few species of bacteria to very high radiation environments in nuclear reactors by developing repair mechanisms for radiation-induced lesions.

Thus if life started on Mars and terrestrial life is a valid analog, then the life could readily adapt to a number of environments that are likely to have occurred on Mars, such as hydrothermal systems, deep aquifers, and transient lakes. A major problem confronting any such life on Mars would, however, be colonization because the surface is so hostile. A strategy of dispersal by the wind and dormancy at the surface, awaiting the occasional moist opportunity, appears likely to fail because of the radiation environment. Subsurface colonization, using the global aquifer system as a dispersal medium, much as the oceans are used on the Earth, is a possible alternative.

The Early Martian Atmosphere

In Chapter 6 we discussed at length the problem of the climate on early Mars and the nature of its atmosphere. Clearly, the chances of life starting on early Mars are greatly enhanced if climatic conditions on early Mars were similar to those on early Earth. The early atmospheres of the terrestrial planets are thought to have been primarily CO_2–H_2O. The CO_2 and H_2O were derived from two main sources: devolatilization during impact and outgassing from the interior. Although much of the carbon was delivered in a reduced form, rapid photolysis of H_2O and loss of hydrogen would have rapidly converted most of the carbon to CO_2. Venus still retains a massive CO_2 atmosphere; most of the Earth's CO_2 is in carbonates. How much CO_2 Mars retained and what early climates were like is a major unresolved issue. On the one hand, there is geologic evidence of high erosion rates and high rates of valley network formation, which suggest a thick early atmosphere and warm surface conditions. On the other hand, there is modeling of impact erosion, which suggests any thick atmosphere would be short-lived, and modeling of CO_2–H_2O atmospheres, which suggests that temperatures could not get above freezing no matter how thick the atmosphere, because of the low energy flux from the early Sun. The lack of detection of large amounts of carbonates is also troubling. The question of what the early atmosphere on Mars was like is thus largely unresolved.

One question that was not addressed in detail in Chapter 6 was the nitrogen content of the early atmosphere. Nitrogen is of interest because of the crucial role it plays in the metabolism of terrestrial life (McKay and Stoker, 1989). The pressure in the martian atmosphere today is 0.2 mbar, or about 4000 times less than at the Earth's surface. Nitrogen in the early martian atmosphere is likely to be in the form of N_2 because of rapid photolysis of other possible nitrogen compounds such as NH_3. McElroy et al. (1977) attempted to reconstruct the history of nitrogen in the martian atmosphere based on the present $^{14}N/^{15}N$ ratio and estimated loss rates from the upper atmosphere. Their models suggest that as much as 50 mbar of N_2 could have been in the primitive atmosphere, but, as discussed in Chapter 6, there are many uncertainties in estimating present loss rates and it is not clear that nitrogen in the present atmosphere retains any record of the primitive atmosphere.

An additional uncertainty in the nitrogen story and one that is of particular interest to biology is the extent to which nitrogen in the atmosphere was fixed as nitrates. Mancinelli and McKay (1988) published theoretical calculations, tied to experimental work of Levine et al. (1982), for the rate of NO production by lightning in a N_2–CO_2 atmosphere. They show, for example, that in a 1 bar CO_2 atmosphere containing 100 mbar N_2, with lightning rates comparable to present Earth, it would require less than 10^9 years to convert half the nitrogen to nitrates. Thus nitrogen could be removed from the atmosphere by two processes, nonthermal escape to space from the upper atmosphere and fixation as nitrates in the regolith.

Incorporation of nitrogen into biologically useful forms is a limiting process for many of the Earth's ecosystems, and its availability may have played a crucial role in the ability of life to start on Mars (McKay and Stoker, 1989). On Earth atmospheric nitrogen is not biologically useful until it is converted into some other form, such as nitrate or ammonia. Bacteria use a nitrogenase enzyme to fix nitrogen. The ability of the nitrogenase enzyme to do this is reduced if the partial pressure falls below 100 mbar, although nitrogen-fixing organisms can still grow, even when the partial pressure falls to a few tens of millibars. Because of the low nitrogen content of the present martian atmosphere and because of its probable low abundance throughout much of martian history, McKay and Stoker (1989) suggested that nitrogen could have played a key role in determining whether life started on Mars and whether it could have survived.

Hydrothermal Systems

Recognition that the most primitive life forms on Earth live in hot springs, that volcanic activity was common throughout martian history, and that the surface of Mars is water rich has led to interest in hydrothermal systems as potential environments in which life might have started or survived. The requirements for formation of hydrothermal systems are (1) suitable rock formations that allow water to circulate to deep levels, (2) a source of heat, (3) an adequate source of water, and (4) a return path to the surface (Ellis and Mahon, 1977). All these requirements are likely to have been frequently satisfied at the martian surface. In a typical terrestrial hydrothermal system, waters circulating at depths of several kilometers become heated and rise convectively. Routes are commonly through faults and fissures. Drilling in the Wairakei field in New Zealand (Elder, 1965) shows, for example, that the high-temperature water rises in a narrow column surrounded by colder water (Figure 8-4). At the surface the hot water moves laterally so that warm water forms a mushroom-like structure a few kilometers wide. Most of the water involved is meteoric water, although there may be small additions from magmatic sources. Water in the rising column is replaced at the base by lateral inflow. Similarly, the outward-moving water at the top of the column cools and rejoins the ambient water system. The turnover times for water in most geothermal systems is on the order of 10^4 to 10^5 years. Frequently a cap rock limits the access of water and steam to the surface, but leakages may create hot springs or fumaroles.

The deep waters are commonly saline with a near-neutral pH. Ca is usually present in signifi-

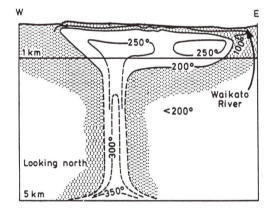

Figure 8-4. Mushroom shaped profile of high-temperature water beneath the Weireki hydrothermal field, New Zealand. The solid lines are measured; the broken lines are inferred. (Modified from Elder, 1965.)

cant quantities, and Si concentrations are usually high. Solutes such as B, Fl, As, NH_3, and H_2S are usually present in higher concentrations than in colder waters. The high mineral concentrations are probably due mostly to reaction between the hot waters at depth and the local rocks, although magmatic contributions occur. The compositions of the waters are controlled by the mineral solution equilibria in the warmer, deeper parts of the system. As the waters rise and cool, deposition occurs. If underground reservoir temperatures are in excess of about 180°C, silica is generally deposited close to the surface, often intermixed and colored with other minerals. Some of the near-surface deposits are remarkably rich in minerals. For example, hydrothermal minerals in open veinlets in the Salton Sea geothermal system include calcite, pyrite, pyrrhotite, sphalerite, and chalcopyrite (Ellis and Mahon, 1977). Scale deposited in wells drilled into the same system contained 10–40% Cu, 4–25% Fe, 7–23% sulfur, 1–6% Ag, 0.5–1% Sb and 0.1–0.2% As. If the deep waters are heated to no more than 100°C, extensive travertine (calcium carbonate) deposits may form. Thus hydrothermal systems are not only warm and wet, but they are rich in minerals and contain a wide range of minor elements. They also incorporate a rich variety of chemical and physical environments in which different strategies for survival of microorganisms might be utilized.

Hydrothermal systems provide two broad classes of environments for life: the warm and wet subsurface, and warm springs at the surface. Any subsurface biology must derive its primary biologic production from the activities of chemolithoautotrophic organisms (Boston et al., 1992). These are microbes capable of fixing inorganic carbon, such as that in CO_2, CO_3^{2-} and HCO^{-3} into living material by means of energy produced from an oxidation reaction of an inorganic compound. Conditions on Mars are probably anaerobic, and Boston et al. list several examples of anaerobic terrestrial organisms that have properties an organism would need to survive in martian hydrothermal systems. Thermophyllic organisms that respire using nitrates have been found growing at 60°C in Icelandic hot springs. Thermophyllic heterotrophs occur in geothermally heated soils near Antarctic volcanoes. Hyperthermophyllic sulfidogenic and methanogenic Archea grow at temperatures between 80 and

110°C around the erupting MacDonald Seamount. Among them are primary-producing H_2 oxidizing species. Huber et al. (1990) found that many of these organisms are related to organisms found within hydrothermal systems off Vulcano, Italy.

Boston et al. (1992) speculated that life could have started on early Mars when surface conditions were more benign; then, as surface conditions changed, organisms could have adapted to survive underground, protected from the damaging UV radiation. But hydrothermal activity is also likely to result in warm, chemically rich surface pools, and such pools were probably common on early Mars when geothermal gradients and rates of volcanism were high. Different strategies that capitalized on the availability of light might have prevailed in these situations.

Possible Martian Habitats

Much of the discussion in this chapter has concerned the origin and limits of terrestrial life, but we do not know whether any of the discussion has any applicability to Mars. Our knowledge of Mars is too rudimentary to know whether environments hospitable to terrestrial life have occurred on Mars or, if they have, where they are. Nor do we know whether limits on present-day terrestrial life represent the outer limits of the conditions under which life could start. We need more information on the range of martian environments so that search for more tangible evidence of life can focus on the most promising locations. The following is a brief discussion of some potential martian habitats based on our current rather meager knowledge.

Ancient hydrothermal and lake deposits

The Noachian and early Hesperian are the times when Mars most likely had warm surface conditions, if we are to judge from the incidence of valley networks and erosion rates. It is also the time when geothermal activity was likely to have been at its highest. Deposits of these ages are therefore natural candidates for biologic scrutiny. But just as on early Earth, most of the Noachian was subject to intense heavy bombardment, and the large impacts would have had the same sterilizing potential as on Earth. It could be argued, therefore, that the best place to look for early life is in the late Noachian or early Hesperian ter-

rains, which formed after the impact rates had fallen but when the heat flow was still high and the remnants of a possible thicker early atmosphere were still retained. Although high rates of volcanism are expected in the Noachian, because of the high heat flows (see Figure 1-14) evidence for volcanism in the Noachian highlands is sparse; the effects of the high impact rates dominate. Nevertheless, several ancient highland volcanoes have been recognized (Figure 8-5). Although direct evidence of hydrothermal activity is lacking, numerous valley networks indicate that water was present, so hydrothermal activity was likely.

Similarly direct evidence for lake deposits is lacking, but they can be inferred from the surface morphology. Most of the valley networks end in local sinks where lakes may have formed. Squyres (1989b) showed several examples where valley networks converge on craters and suggested that lakes formerly formed within the craters. In addition, several features within the highlands that have unusually high albedos have been interpreted as playas (Lee, 1993).

Lakes within the canyons

The possibility of long-lived lakes within the canyons was discussed in Chapter 3. The evidence is stacks of layered sediments several kilometers thick in various parts of the canyon (see Figures 3-4 and 3-11). The deposits appear to be upper Hesperian in age (Scott and Tanaka, 1986), but this is the age of the upper surface. The extreme thickness of the deposits suggests that they were accumulating for a long time, although we have no way of dating the start of deposition.

On the assumption that the lakes initially formed when surface temperatures permitted liquid water at the surface, McKay and Davis (1991) attempted to calculate how long the lakes would outlast a change in climate. They used the climate model of Pollack et al. (1987) and employed calculations similar to those used to explain the survival of ice-covered lakes in subzero conditions in the Antarctic (see Figure 3-13). As discussed in Chapter 3, lakes in the dry valleys of Antarctica do not freeze solid, despite the low temperatures, because for a few days a year temperatures rise above freezing and the lakes receive small influxes of meltwater. This provides

Figure 8-5. Ancient volcano in the Noachian uplands at 20°S, 187°W. The summit caldera is 8 km across. Such volcanoes may have hydrothermal deposits associated with them (Viking Orbiter frame 430S23).

enough heat to offset losses from the ice on the lake's surface. The ice thickness is maintained at 3–6 m as ice lost by ablation from the upper surface is replaced by freezing of new ice onto the lower surface. In applying this model to Mars, McKay and Davis concluded that if there was a source of meltwater, liquid water habitats could have been maintained under thin ice covers for several hundred million years after mean global

temperatures fell below freezing. At that point, the summertime peak temperatures fell below freezing, cutting off the supply of meltwater. However, as discussed in Chapter 4, water is more plausibly supplied by the groundwater. If there was groundwater in the surrounding terrain, as is very probable, it would leak into the canyon lakes to maintain the level within the lake the same as that within the local aquifers (assuming that movement is slow so that the groundwater surface is essentially flat). This does not change the conclusions of McKay and Davis, but instead emphasizes that the canyon lakes could have been extremely long lived.

An additional aspect of the Antarctic lakes of biologic interest is that atmospheric gases become concentrated within the waters of the lakes. The meltwaters bring in dissolved gases, and these accumulate within the lake, as gases are excluded from the ice as it freezes onto the bottom of the ice cover. A similar process would occur on Mars, irrespective of whether the inflowing water was meltwater or groundwater. McKay and Davis suggest that this could be a significant source of biologically important gases as the martian atmosphere thinned.

Post-Noachin volcanic regions

One of the main themes of this chapter has been that if life started on Mars, then hydrothermal deposits are among the most likely places for it to have started or survived. Given that volcanism was common and water was widespread, hydrothermal activity must have been common, and we have supporting evidence for circulation of warm fluids from SNC meteorites (Watson et al., 1994). Unfortunately, we do not know where to go to find hydrothermal deposits; we have not yet made the observations necessary for their detection. The results of hydrothermal activity would be difficult to recognize in imaging from orbit. Hydrothermal deposits would be most easily recognized by their unique mineralogy, for which we need high-spatial and high-wavelength resolution spectral mapping. Hydrothermal activity occurs at vents or where there is some intrusion at depths, so we need some way of identifying these locations. It serves little purpose to search for life in volcanic deposits, such as a lava plain, far from a vent. Even on Earth, life has difficulty colonizing lavas.

A number of martian surface features have been attributed to volcano-ice interactions, as discussed in Chapter 5. These include mountains that form by eruptions from a central vent under ice and moberg ridges that form in a similar manner but from fissure eruptions. Allen (1979) and Squyres et al. (1987) identified numerous possible examples, especially in Elysium. The apparent initiation of outflow channels (see Figure 5.17), and smaller fluvial-like features in the vicinity of volcanoes (Mouginis-Mark, 1985) appear to be clear evidence of volcano-ice or volcano-water interaction. Dissection of the surfaces of some volcanoes (see Chapter 4) is additional evidence of volcanic activity stimulating the movement of water. However, most of these examples have little to do with hydrothermal activity. They result from discrete eruptions of volcanic rocks into water-rich or ice-rich ground. They are not the result of sustained groundwater circulation, except possibly for the dense arrays of valleys on some of the volcanoes. Identification of plausible volcanic habitats on Mars must await more information.

High-latitude water-rich deposits

Stability relations and geomorphic evidence (see Chapters 2 and 5) indicate that ice should be pervasive in near-surface materials at high latitudes. One reason a high-latitude site was chosen for the *Viking 2* lander was a desire to go where water (ice) would likely be present. During periods of high obliquity, the transient presence of liquid water may be quite common at high latitudes. Heating rates are so high that sublimation from an ice deposit at the surface may be insufficient to dissipate the heat and melting results, temporarily providing liquid water. Water could also exist transiently in ice-rich soils as they are heated. Humidity within the pores would inhibit sublimation, and melting would ensue if temperatures rose above 273 K, as commonly occurs. Accordingly, some biologists favor searches at high-latitude sites.

These four types of sites do not exhaust the possibilities. As indicated earlier, at present we do not have enough information about the martian surface to intelligently decide where to focus the search for biologic activity. We will be in a much better position to do this once we have accomplished global mineralogic and chemical surveys.

Summary

The surface of Mars at present is very hostile to life. The high UV flux, the lack of liquid water, and the oxidizing character of the soil all militate against survival of any living forms on the surface. But conditions on early Mars may have

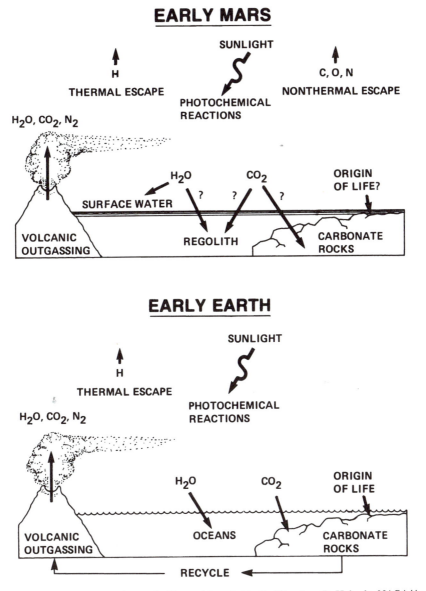

Figure 8-6. Diagram comparing early Mars with early Earth. (Courtesy R. Haberle, NASA/Ames.)

been very different. By 3.5 Gyr ago life on Earth was already diverse and complex. We do not know the environment under which life started on Earth, but hydrothermal springs and volcano-tidal interfaces are likely candidates. Early Earth and early Mars were in many ways similar (Figure 8-6). Both had high rates of impact, high rates of volcanism, abundant water, and a N_2–H_2O–CO_2 atmosphere. Both may have had warm climates, although this is not absolutely certain for Mars. We know life started on Earth, but we do not know what happened on Mars. Surface conditions on Mars changed around 3.5–3.8 Gyr ago, as indicated by changes in the rates of erosion and valley formation. If life did start, it would have had to adapt to the changes, and subsequent colonization of possibly rare and transient habitats would have been hindered by the harsh radiation environment at the surface. The best places to look for past life are probably in lacustrine sediments and hydrothermal deposits, which should both be common at the martian surface. Even if life never started on Mars, the search in these favored locations may provide clues concerning prebioitic chemical evolution.

9 FUTURE MARS EXPLORATION

Missions are the lifeblood of planetary science. Unfortunately, planetary missions are expensive and because they are expensive only a few nations or agencies have independent planetary exploration programs. At the time of this writing the list was restricted to four: the United States, Russia, Japan, and the European Space Agency. To lay out a rational, step-by-step program for the exploration of Mars is not difficult, but the likelihood of such a plan being followed is small. Because the missions are expensive, their approval requires support of a much broader constituency than the small planetary science community. Planetary missions may serve a variety of purposes in addition to advancing scientific knowledge. They may be used to demonstrate to the rest of the world a nation's technical virtuosity and the superiority of its economic system. They have been used to promote technological innovation and to preserve specialized technical capabilities. They may foster international cooperation and cement economic alliances. They stimulate interest in science and technology among a nation's youth. They must also have public appeal, for it is the public who pays for them. This multiplicity of objectives often leads to frustration in planning, for not only do the different goals pull the exploration strategies in different directions, but the weight given to the different goals is constantly shifting with the political climate.

In recent years, because of economic conditions or changes in national priorities, it has become increasingly difficult for individual nations or agencies to sustain their own independent planetary programs. Pressure has therefore mounted for all the participating countries to cooperate in a unified international exploration program, with each country participating according to its unique desires and capabilities. This is particularly true with respect to Mars, for interest in Mars exploration appears universal among those nations with ambitions in space. Cooperation may take place at a variety of levels of interdependence. At the lowest level of cooperation, different agencies simply agree to fly complementary missions. At the next level, one nation may agree to provide instruments or components for another's mission, in such a way that failure to deliver does not jeopardize the entire mission. At the highest level of cooperation are truly joint missions in which essential subsystems are provided by different agencies, and failure to deliver causes failure to the mission. All those modes of cooperation have been and are being followed.

All space agencies periodically revise their strategic plans in light of the current fiscal and political realities. Different elements of the programs are re-examined and resources allocated according to their availability and the current priorities. For planetary missions the process is complicated as a considerable amount of work must be done on any proposed mission before its technical feasibility and costs are understood. Proposed missions, therefore, go through several levels of study and approval. Many missions may be studied in a general way to get a rough assessment of feasibility and cost, but in order to go on and do the work necessary to achieve the next level of understanding, missions must compete against each other to narrow the field of those to be studied in greater depth. Many missions are conceived but few are flown. In recent years it has typically taken almost a decade from conception to launch. These long planning cycles, while they help to rationalize choices between competing missions and protect against unforeseen cost, schedule, and technical difficulties during implementation, do add to the program costs. There is thus a continual tension between the desire to cut planning costs and the desire to commit only to missions that we know can be realized within the estimated budgets. The long planning cycles have international implications, because every country and agency has its own unique political process for approval and its own unique funding cycle. In order to cooperate internationally, ways must be found to coordinate across these differences.

At the time of writing, a philosophical change was underway in the United States concerning how best to do planetary missions. Emphasis was shifting from large, infrequent, expensive missions with comprehensive science goals to smaller, more frequent, less expensive missions with highly focused goals. In order to gain support of a wide scientific constituency, missions in the past have tended to address a broad range of science objectives, thereby imposing on the missions a high level of complexity and cost. Their complexity and cost emphasized the need to prove feasibility in the planning stages, and stimulated concerns about technical risks and cost overruns during implementation. Numerous planning cycles and technical and fiscal oversight procedures were therefore mandated, further adding to the costs. As costs escalated so did concerns of failure. The result was a somewhat unstable situation in which cost, risk, and complexity drove each other upward. With smaller but more frequent missions, many of these escalating factors can be avoided. The simplicity of the msisions reduces the need for long planning cycles and burdensome oversight and integration procedures. Multiple launches distribute the risk so that a single failure is not so catastrophic, and individual missions can be tailored to unique requirements of particular science objective. However, while these are very obvious advantages, the ability of a series of small simple missions to achieve complex and comprehensive science goals remains to be demonstrated.

This book is about Mars science, and this chapter looks at Mars exploration primarily from the scientific viewpoint. Despite the many different reasons for engaging in planetary exploration, individual missions must have a sound scientific rationale. We will look at the scientific goals of Mars exploration and the mix of missions and measurements that might address those goals. Although the focus of this book is water, the discussion of the exploration goals is more general, including several issues not closely related to the water story.

One goal of Mars exploration that has scientific implications but that is not necessarily an essential part of a scientific strategy is to prepare the way for eventual human exploration. Mars is surely the first planet away from the Earth-Moon system to which humans will go, and it seems inevitable that one day people will go there. The rationale for human exploration is not scientific. People will go to Mars not because of a desire to see if there is indigenous life or to better understand climate change. They will go because of a compulsion to explore, because of the challenge it presents to humankind's will and ingenuity. Other reasons will no doubt be listed: energizing the economy, stimulating education, providing technological spin-offs, sustaining aerospace capabilities, and stimulating national pride. But I believe that the drive to explore is mostly spiritual, not practical. Exploration lifts us above the humdrum concerns of food and shelter, and provides us with feelings of awe, wonderment, and pride. It is this aspect of exploration that I think I find most compelling and the one that I think will ultimately drive us on to Mars.

The prospect of eventual human exploration presents Mars exploration with near-term goals in addition to the scientific and political goals outlined earlier. We must examine what knowledge of the planet is needed to deliver people to Mars and return them safely to Earth. This would include knowledge of immediate engineering concern, such as how easy it is to land and move around the planet, and what resources are available for fuel, building, and sustaining life. But we should also be concerned about broader scientific issues. If people are to do useful work on Mars, they must go to the right places, with the right tools, and the right science objectives. We should therefore prepare the way with the appropriate missions. We should perhaps also be conscious of what science goals might best be achieved by human exploration and so defer addressing those goals until they become a reality.

Science Goals

In the broadest sense, the scientific goals of Mars exploration are to determine the present state of the planet, how it formed and evolved to that state, and how that evolutionary history compares with other planets, including the Earth. We have seen that Mars has had a long and varied history, and is extremely rich in the science questions that it poses, touching as they do on geology, geophysics, climatology, biology, and aeronomy. The scientific goals can be consolidated into three broad, interconnected themes: (1) to determine the mode of formation and evolution of the solid planet, (2) to determine the planet's climatic history, and (3) to determine its biologic history. The following few paragraphs outline the main goals in these three areas. For more comprehensive accounts of strategic goals for the scientific exploration of Mars see COMPLEX (1978), Mars Science Working Group (1991), and International Mars Exploration Working Group (1995).

Mode of formation and evolution of the solid planet

In Chapter 7 we saw how important the formation of the planet and very early events are for the story of water on Mars. Key questions with respect to the formation of Mars are as follows: From what materials and in what sequence did the planet accumulate? How quickly did the planet accrete? What was the state of the interior and the state of the surface and atmosphere at the end of accretion? What is the present internal structure and how did it arrive in that state? To address these questions requires a broad combination of geophysical, geochemical, and geologic measurements. The composition of the core and mantle will tell us much about the materials from which the planet accumulated. The composition of the core can be derived from its size, determined from seismic measurements, from the history of the magnetic field as preserved in ancient rocks and from the composition of the mantle. Mantle composition can be inferred indirectly from volcanic rocks or measured directly in pieces of mantle brought to the surface in the volcanics. Mantle compositions, particularly isotopic compositions, also provide information on the state of the mantle at the end of accretion and how it has since evolved. We need to get compositional information in different places because the mantle is likely to be laterally inhomogeneous, as evidenced by sustained volcanism in local regions such as Tharsis and Elysium. Gravity and altimetry will provide information on variations in the thickness of the crust and lithosphere, and how they have changed with time.

Understanding the surface is crucial for understanding the planet's evolution. Most geologic events leave behind some kind of record and partly destroy the record of previous events. On Earth the relatively fast pace of geologic activity has destroyed much of the evidence of ancient events, but on Mars the geologic record spans the planet's entire history. The record is partly in the morphology of the surface and partly in the chemistry, mineralogy, lithology, and three-dimensional configuration of the surface rocks. The task of the geologist is to reconstruct the history of the planet from this fragmentary record. Age is a fundamental part of the story. Not only must the events be identified from the clues that they leave behind, but the time that the event occurred must also be determined. We wish to know the sequence and nature of the main volcanic and tectonic events, and how the history of erosion and deposition is related to these events. Of particular importance, of course, are events involving water. What was the initial inventory of water on the planet and how has it evolved with time? Where is the water now? How, when, and under what conditions did the various water-worn features form? Were there ever ocean-scale bodies of water on Mars, and if so, how did they form and dissipate?

Climate change

To better understand climate change we must better understand the nature of potentially climate-sensitive processes, such as valley formation and weathering, and place them in historical context. Thus, elucidation of the climate history is intimately tied to understanding the general geologic history. Climatically important processes include formation of lakes, valleys, flood features, debris flows, fretted terrain, and a wide variety of seemingly glacial and periglacial features. The list also includes precipitation of carbonates from the atmosphere, fixation of nitrogen in the soil, and possible trapping of water at high latitudes. Detection of thick carbonate deposits would support the supposition that Mars had a thicker atmosphere in the past. Samples of past atmospheres trapped in surface materials would provide very direct evidence of past climates. Evidence would also be provided by isotopics of volatile species, such as nitrogen, carbon, oxygen, and the noble gases, not only in the atmosphere but also in different surface materials. The isotopic evidence is particularly important for elucidating events very early in the planet's history for which other records have been destroyed. Weathering horizons in soils and the mineralogy of weathered products will provide information on past conditions and how they have changed with time. Much of the liquid water involved in processes at the surface must have ultimately been deposited in low-lying areas such as the northern plains and large impact basins. Exploration of these areas and their deposits would resolve many of the issues concerning the fate of water and the possibility of massive episodic climate changes. The best record of more recent, and possibly more modest, climatic fluctuations is in the layered deposits at the poles, so these are also of interest.

Understanding the present atmosphere also has climatic implications, because knowledge of how the present atmosphere works will enable us to extrapolate to other times or other circumstances, such as when thicker atmospheres might have been present or obliquities were different. Losses from the upper atmosphere are another important part of the climate story, for not only may significant losses of different volatiles have occurred, but the losses have left a time-dependent record in the isotopic gases that remained. In order to understand how the atmosphere has evolved, we need to determine present loss rates, particularly of H, C, N, and to O, and to understand better the loss mechanisms.

Biologic history

The harsh conditions at the present surface suggest that the chances that life could survive on present-day Mars are very small. However, evidence of abundant liquid water at the surface in the past indicates that at times conditions may have formerly been much more hospitable. Early objectives for biology are to understand better the climatic history of the planet and to locate materials, such as lacustrine or hydrothermal deposits, that are indicative of past water-rich environments. These materials would then be examined for biologically induced chemical and isotopic changes, and for more direct evidence for life, such as organic remains. If it were found that life had indeed started, then we would need to search for and explore specific favorable niches, such as subsurface aquifers or active hydrothermal vents. Definitive detection of past life will likely require samples here on Earth.

Precursor Knowledge for Human Missions

Two very different types of precursor knowledge are needed to undertake human missions: (1) detailed knowledge of engineering characteristics of potential landing sites and their environs, and (2) knowledge of the scientific potential of landing sites and the specific science questions that can be addressed at those sites. While human missions may be decades away, some of the needs can be anticipated.

The engineering needs can be conveniently classed into three areas of concern: safety, habitability, and resources. If aerobraking or aerocapture are to be used for landing or orbit insertion, then more detailed knowledge of wind shear and other conditions in the upper atmosphere will likely be needed. Knowledge of the wind characteristics in the lower atmosphere would also be desirable for potential landing sites, irrespective of the landing system used. Although the landers will likely be robust, information on block and slope frequencies will surely be needed in order to confirm that the site characteristics fall within the acceptable ranges. Surface elevations will need to be accurately known, as well as characteristics of the local terrain. The physical and chemical properties of the ubiquitous dust will be of concern for the potential effects on machines and life support systems. Fortunately, many environmental factors, such as surface temperatures, dust storm frequencies, and radiation levels, are already quite well known.

Human missions will probably involve long stays on the martian surface. Sites with thick regoliths will be favored because the regolith can provide materials that can be used to provide protection against radiation. Lava tubes have also been suggested as potential shelters. The site must be trafficable for both machines and people.

Detailed knowledge will therefore be needed of the physical properties of different surface materials and their distribution about potential sites. The atmosphere is one resource that is everywhere available, and pilot systems have already been built that could be used on Mars to extract oxygen from CO_2 using solar power. But many other potential resources may be available. These include water-ice and water of hydration for extracting both hydrogen and oxygen, and carbonates, nitrates, sulfates, and other salts.

The scientific knowledge required for human missions is similar to that required for purely scientific exploration. In order to determine how people can most effectively do science on Mars we need to know what the crucial questions are, what measurements need to be done to resolve those questions, and where to go to do the measurements. Mars is far more complex than the Moon, and our knowledge of Mars is more advanced than our knowledge of the Moon was at the time of Apollo. Most of the science learned on the Moon came from returned samples that were chosen with little scientific discrimination. But such a strategy is unlikely to be successful in resolving important climatologic and biologic issues on Mars. To take advantage of the unique capability of people, we should choose our landing sites very carefully. One might, for example, have a successful mission to the *Viking 1* landing site or to the plains of Tharsis, yet have a scientific failure because the critical questions concerning climate and life could not be addressed. If people are to do useful science on Mars, human missions should be preceded by a scientifically driven program of precursor missions that enable us to define the science questions, to discriminate between those tasks best done robotically and those best done by people, and to evaluate the potential of different sites for pursuing those tasks best done by people.

One unique talent that people would bring is the ability to do biologic and geologic field work. Field work is a complicated process in which the person on the ground is constantly making almost instantaneous decisions on where to look, where to go, where to stop, and when to backtrack. These successive decisions are usually based on knowledge and experience, gained over years of training and practice. The ability to draw inferences from various clues at an outcrop, based on clues at other scales such as the regional geology, or some microstructure at a very different outcrop several years ago, is not something that can likely be taught to a robot. This activity is extraordinarily difficult to do robotically on Mars if the humans operating the robots are here on Earth because the roundtrip signal times are tens of minutes. It is the type of activity that has led to discovery of early life on the Earth and may be required to do effective searches on Mars. But these unique capabilities of trained people can only be used effectively on Mars if they go to the right places.

Exploration Approaches

Mars exploration may be conducted by many different types of missions: remote sensing, independent landers, networked landers, penetrators, rovers, balloons, sample return, and human missions. Each technique is different in its suitability for addressing different science objectives, and an ideal exploration program would ultimately include a mix of many different components. The following paragraphs outline general capabilities of different techniques and indicate how the technique might fit into a general exploration plan.

Orbital remote sensing

Imaging from orbit forms the basis of much of what we currently understand about the geologic evolution of Mars. Different geologic processes affect terrain at different scales, and experience shows that objects can be located in different scale images only when scales differ by factors of 10 or less. Viking has provided us with almost global coverage at roughly 200 m resolution and samples of small areas with resolutions down to 10 m. For geologic purposes, we need to acquire broad regional coverage of different areas of the planet at 10–20 m resolution and small areas, globally distributed, at 1–4 m resolution, so that all scales are examined and imaging at each scale can be placed in context of coarser scale imaging. In addition to this high-resolution imaging, meteorologists are interested in monitoring global weather from orbit at resolutions of several kilometers at the same time that meteorology stations are operating at the surface.

At the time of this writing we had almost no information on chemical and mineralogical variations in the surface materials. Viking and Phobos data, and telescopic spectral data, are all consistent with the surface rocks being dominantly basaltic and the soils being partly hydrated, poorly crystalline products of weathering, but we have very little specific information on rock and soil chemistry, and no information on how the mineralogy and chemistry vary according to location. Gamma-ray spectroscopy is the best technique for mapping chemical variations. Mineralogic variations may be determined by visible and near-infrared reflectance spectroscopy and by thermal emission spectroscopy. Only experience will show which is most suitable for Mars. Neither technique will likely lead to definitive identification of the mix of minerals at specific surface locations, but both techniques will enable us to distinguish differences from area to area and place constraints on the causes of the differences. From a strategic viewpoint, additional high-resolution imaging and mapping of chemical and mineralogic variations should be achieved prior to placing many landers on the surface so that the landers can be placed to sample the maximum geologic variety.

Altimetry and gravity are crucial for understanding the density variations, thermal state, and the state of stress of the crust and upper mantle. The measurements can only be made globally from orbit. To acquire the required precision, an orbiter carrying an altimeter, and an ultra-stable oscillator (for tracking) must be placed in a low circular orbit. Surface elevations are currently known only to within 1–2 km and possibly worse (Figure 1-8). If parachutes are to be used to decelerate landers, then these errors must be taken into account and the landers designed to land at higher than actual elevations, with resulting sacrifice of payload mass. We need, therefore, to accurately determine elevations before embarking on an ambitious lander program. Orbiters can also be used for sounding the surface at radar wavelengths to determine variations in density, permittivity, and conductivity to depths below the surface of a few hundred meters. The main interest is detection of water and ice.

In order to understand the circulation of the at-mosphere and what drives it, we need to measure how temperature, pressure, and aerosol profiles in the atmosphere change globally by time of day, season, and interannually. A variety of techniques can be used from orbit to obtain such profiles, including pressure-modulated infrared radiometry, thermal infrared radiometery, and radar techniques. A strong desire is to have these measurements performed at the same time that meteorology measurements are made at the surface.

Characterization of the interaction of the planet with the solar wind can only be done from orbit and is important for understanding losses of volatiles from the upper atmosphere. A wide variety of measurements are needed to characterize the solar wind itself, how it interacts with the upper atmosphere, and what gases are irretrievably lost from the planet downstream.

Simple landers

Landers can be used as independent stations or in combination with other landers and/or orbiters. Some science objectives, including most of those for geology and geochemistry, can be achieved with independent landers, with each successive lander adding incrementally to our knowledge. Other objectives, including most of those for seismology and meteorology, require simultaneous measurements from an array of landers. A variety of lander approaches can be identified. Each approach differs according to its suitability for addressing various science issues. They can be sequenced and combined in various permutations according to programmatic opportunities to create a coherent lander program.

GEOSCIENCE STATIONS. The main function of a geoscience station is to characterize the geology at a particular point on the surface. Characterization could include imaging; determination of the lithology, chemistry, mineralogy, and isotopics of the rocks and soils; and determination of the physical properties of local materials. The stations could include a small rover to image beyond the immediate lander area and to deploy instruments to measure soil and rock chemistry. Such stations need not be long lived because their objectives are accomplished once the site has been characterized; nor do they need to operate simultaneously with other stations. Data rate requirements are modest. Operation of some stations at high latitude, where water-ice is stable at shallow depths, is desirable. Similar stations could also be used to determine more precisely the chemical and isotopic composition of the atmosphere.

SEISMIC NETWORK. The main objective of a seismic network is to determine the internal structure of the planet. To do this at least four, but preferably several, sensitive seismometers must operate together on Mars for a period long enough (years) to ensure that several marsquakes are detected by a few stations simultaneously. It may be desirable first to make pilot measurements to determine the seismicity of the planet and the characteristics of seismic signals on Mars before emplacing a network. The main characteristics of the seismic stations are simplicity in instrumentation, in addition to long life, high data transmission rate, and simultaneity of operation with other stations. A communications orbiter may be required because of the high data rates. While the prime function of a seismic network is to determine the internal structure of the planet, the network may also have the capability of detecting present-day volcanism and so be useful for focusing the search for warm, wet places.

METEOROLOGY STATIONS. Meteorology stations could be very simple. The highest priority instruments, those that measure pressure, temperature, wind, and humidity, are simple and lightweight. The data rates are low, but the stations must last for at least one Mars year. After early missions have completed pilot measurements, a meteorology network is desired. In order to characterize both meridional and zonal flow, such a network should have at least 10–20 stations, with some at high latitudes, and simultaneous sounding from orbit. The orbital measurements define the baroclinic component of the circulation (the component that depends on height). The surface measurements measure the barotrophic component (the component that is independent of height).

PENETRATORS. Penetrators are dartlike objects dropped from high altitude to stick into the ground. They have the virtue of getting instruments below thermal and radiation effects at the surface. They can also provide efficient coupling to the ground for seismometers and enable the placement of analytic instruments, both chemical and thermal, directly against local materials. However, penetrators are inherently more risky than other types of landers because of the strong shocks they sustain on landing. Thermal problems also constrain lifetimes. Penetrators are suitable for measuring soil chemistry, heat flow, and subsurface water, but their restricted lifetimes do not make them very suitable as seismic and weather stations.

BALLOONS. Balloons can be used both as instrument platforms and to deliver small landed packages to the surface. They can be designed to float at specified altitudes and to approach the ground periodically to drag instruments across the surface in a guide rope. Remote sensing instruments in a gondola could include a camera, magnetometer, radiometer, and various meteorology sensors. A guide rope could include ground-penetrating radar, and chemical and mechanical sensors. One obvious advantage of balloons is their mobility, which permits high-resolution profiles across distances, possibly as large as thousands of kilometers. Possible disadvantages are small payload masses, difficulty in controlling the flight path, and short lifetimes.

POLAR SOUNDER. The poles are unique in that they appear to preserve records of climatic variations and exchange of volatiles between the surface and atmosphere in the geologically recent past. The main science objectives of a polar lander would be to make measurement profiles through this record, as preserved in the layered terrains. The profiles could be a combination of remote sensing and direct measurements. Polar landers present unique implementation problems because of hostile winter conditions and poor direct-to-Earth communications.

Advanced missions

LONG-RANGE ROVERS. We have already discussed the desire for local rovers to explore and deploy instruments in the immediate vicinity of simple geoscience landers. Rovers have several potential functions, and their design and capability might vary according to which functions are emphasized. They extend the reach of landed missions, enabling analyses and observations to be made where landings may not be possible.

Rovers can be used to collect materials either locally or over long distances for analysis at some parent vehicle or for analysis back on Earth. They can carry instruments to perform in situ analyses on a variety of rocks and soils. They can emplace geophysical stations and, while on traverses, perform geophysical measurements, such as active seismometry and electromagnetic sounding to detect water. They can make engineering tests of surface materials in anticipation of more sophisticated missions.

However, the most outstanding attribute of capable, long-range rovers is that they perform exploration, and probably no other type of mission, other than human missions, can so effectively convey the excitement of exploration as a rover that can travel tens to hundreds of kilometers through different types of terrain.

MARS SAMPLE RETURN. Sample return from Mars has long been a goal of planetary exploration. Many critical measurements are simply too complicated and interactive to be performed remotely, at a distant planet, in the foreseeable future. Most age-dating techniques, for example, stretch the capabilities of terrestrial laboratories. In most geologic processes, trace elements are more strongly fractionated than the major rock-forming elements, and so are more sensitive tracers of geologic history. Many critical elements, such as iridium and europium, are found only at parts per million or parts per billion concentrations, and are very difficult to determine at the appropriate sensitivity, even on Earth. Microscopic techniques for recognition of Archean life forms on Earth have been developed only in the last two decades and would be extremely difficult to automate. Thus, many crucial measurements require samples here on Earth. Moreover, it is impossible to know a priori which are the most critical measurements to make. Having samples here on Earth allows the measurement strategy to shift in response to previous analytic results, and new techniques can be developed, as suggested by the samples themselves.

Experience with the Moon emphasizes the enormous power of returned samples when placed in the context of global data. Sample data are the basis for almost all our current ideas about the Moon. Having lunar samples in hand allowed the complete analytic and intellectual capacity of the science community to focus on the Moon's evolution. Instead of a small, predetermined set of analytic techniques applied to in situ samples, acquisition of lunar samples enabled the approach to be all-encompassing and flexible, with the analytic emphasis shifting as the meaning of each set of results became better appreciated. There is no reason to believe that these enormous advantages would be any less for Mars. Indeed, the apparently more complex geology of Mars and the biologic potential enhance the importance of the kind of comprehensive examination that returned samples allow.

The availability of samples here on Earth, in the form of SNC meteorites, places unique requirements on Mars sample return. For the Moon at the time of Apollo, return of any sample would have inevitably led to major advances in our understanding, but this is not true for Mars. Return of more relatively young basalts from Mars may advance our knowledge only incrementally. We need samples of biologic or climatologic interest, such as ancient fluvial deposits, lacustrine sediments, or hydrothermal deposits. These will be only limitedly accessible, so careful search for appropriate sites should be completed before committing to the expense of sample return.

Another strategic issue with respect to sample return is mission complexity. Two fundamentally different approaches may be taken. One approach is to have few, relatively large and complex missions; the other to have several simple missions. The large missions would emplace a substantial capability at a few sites. Included would be a long-range rover that could travel tens or hundreds of kilometers to collect samples representative of the region. These could amount to several kilograms and would be assembled and packaged at the landing site for delivery back to Earth. The second approach would be to send a limited capability to several sites. These landers would have a very simple rover that could travel only tens of meters from the lander to collect a few hundred grams for delivery back to Earth. Geologic complexity would be a desirable attribute of landing sites in the first approach and geologic simplicity in the second approach.

The cost of sample return can be significantly reduced by utilizing resources at Mars. One promising approach is to extract oxygen from

CO_2 in the atmosphere. Because oxidants do not need to be delivered to the martian surface, the lander size is significantly reduced.

Future Prospects

The logical way to devise a strategy is to prioritize one's goals, to define what knowledge is needed to achieve those goals, to decide what instruments and capabilities are needed to acquire that knowledge, and then to integrate the instruments and capabilities into a mission. This path has been followed in the past. The Viking mission, for example, had as its main goal the search for life, and the mission was specifically designed to meet that goal. Complex instruments were built for life detection and organic analyis, and stringent requirements on sterilization and contamination were met. But the Viking model is unlikely to be followed for future missions to Mars in the next decade. Cost will likely be the over-riding factor. Missions will be designed not to do what is necessary to achieve a specific goal but to do the best possible work within the cost constraints. The challenge will be to do credible science with missions that are budgeted at the equivalent of a factor of 10 lower than those of the past.

The cost problem is exacerbated by the current stage of Mars exploration. The initial reconnaissance was achieved in the 1960s by *Mariners 4, 6, and 7.* Global characterization was started by *Mariner 9,* continued with the Viking and Phobos missions, and will be largely completed before the end of the century, by which time missions currently being implemented will have been flown. After that time attention will shift to the surface. Lander missions are intrinsically more expensive than orbiter missions. Moreover, stationary landers are useful for only a few purposes, such as components of meteorology and seismic networks. To explore the surface, mobility is required, for which rovers are needed, which further adds to the cost.

At the time of writing, the United States had approved a long-term Mars exploration program but with extremely tight cost guidelines. Better understanding of the water story had been identified as the main goal of the program. Two missions were being implemented: Pathfinder and Mars Global Surveoyor. Mars Pathfinder was largely an engineering experiment to devise a low-cost system for landing on Mars and to explore ways of streamlining implementation procedures for planetary missions so that they could be accomplished with major reductions in cost compared with previous missions. It was to carry a camera, a meteorology station, an Alpha-Proton-X-ray spectrometer (APX) to analyze local rocks and soils, and a small rover. Mars Global Surveyor, to be launched in 1996, was to do most of the global mapping from orbit originally planned for the ill-fated Mars Observer. The mapping was to include high- and moderate-resolution imaging, near- and far-infrared spectrometery, altimetry, and mapping of the gravity and magnetic fields. U.S. missions beyond Mars Global Surveyor were to include an orbiter that might map the chemistry of the surface with a gamma-ray spectrometer and/or do sounding of the atmosphere with an infrared radiometer. Attention was then to shift to landed science.

The Russians also intended to launch a mission in 1996. This is an ambitious mission, with an orbiter, two small landers, and two penetrators. The orbiter included cameras and spectral instruments that complemented those of Global Surveyor. The landers and penetrators were to carry cameras and a variety of geochemical, geophysical, and meteorological instruments. At the time of writing, the Japanese planned to launch a mission, called Planet-B, to investigate interactions of the solar wind with the upper atmosphere, and the European Space Agency was looking into the feasibility of launching several small landers to the surface in 2003.

Completion of the global remote sensing is essential before embarking on an extended lander program. This may not be important for some lander goals. Meteorology or seismologic goals may, for example, be successfully accomplished, even if site characteristics are poorly known before landing. But to address the main concerns of this book, which are the history of water, climate change, and the possibility of life, going to the right places with landers is crucial, and orbital observations will provide the information needed to make the choices. It serves little purpose to de-

Table 9-1. Observations needed to improve our understanding of the inventory and distribution of water

Knowledge needed	Technique	Goals
Global terrain, chemistry, mineralogy. (orbiter)	Imaging VIS-IR spectroscopy γ-ray spectroscopy	Understand water processes. Identify former oceans, lakes, springs. Define erosion rates vs. time.
Atmosphere istotopics - D/H C, N, O, noble gases.	Mass spectrometry	Understand water history - original amounts, losses, and gains.
Microfeatures, in-situ chemistry, mineralogy. (lander with rover)	Micro-imaging APX[1] DSC/EGA[2]	Understand water processes - sediments precipitates, etc. Depositional conditions. Weathering conditions.
Geophys./geochem. profiles. (lander with rover)	γ-ray spectrometery n-spectrometry[3] EM sounding[4]	Water at surface Water near surface (<1-2 m) Deeper water.
Soil/rock isotopes. (lander with rover)	Mass spectrometry	Age of weathering. Composition of past atmospheres.
Definitive trace elements, lithologies, ages, etc.	Sample return	Environmental conditions, fractionation processes, ages, etc.

[1]Alph-proton-X-ray spectrometer [2]Differential scanning calorimeter/evolved gas analyzer [3]Neutron spectrometer [4]Electro-magnetic sounding.

Table 9-2. Observations needed to improve our understanding of climate change

Knowledge needed	Technique	Goals
Global terrain, chemistry, mineralogy. (orbiter)	Imaging VIS-IR spectroscopy γ-ray spectroscopy	Understand water processes. Identify former oceans, lakes, springs. Define erosion rates vs. time.
Atmospheric isotopes - D/H C, N, O, noble gases. (orbiter)	Mass spectrometry	Understand water history - original amounts, losses, and gains.
Atmospheric sounding (orbiter)	IR Radiometry Radar	Determine structure and circulation of atmosphere.
Loss rates of gases from upper atmosphere. (aeronomy orbiter)	Many interrelated	Understand atmosphere depletion.
Microfeatures, in-situ chemistry, mineralogy (lander with rover)	Micro-imaging APX DSC/EGA	Understand water processes (sediments, precipitates, etc.). Depositional conditions. Weathering conditions.
Soil/rock isotopes. (lander with rover)	Mass spectrometry	Age of weathering. Composition of past atmospheres.
Polar lander.	Radar sounding, drilling	Quasi-periodic climate change.
Definitive trace elements, lithologies, ages, etc.	Sample return	Environmental conditions, fractionation processes, ages, etc.

Table 9-3. Observations needed to better assess the possibility of indigenous life

Knowledge needed	Technique	Goals
Global terrain, chemistry, mineralogy. (orbiter)	Imaging VIS-IR spectroscopy γ-ray spectroscopy	Understand water procceses. Identify former oceans, lakes, springs. Define erosion rates vs. time.
Atmosphere isotopics - D/H C, N, O, noble gases.	Mass spectrometry	Understand water history - original amounts, losses, and gains.
Microfeatures, in-situ chemistry, mineralogy. (lander with rover)	Micro-imaging APX DSC/EGA MOX[1] GCMS[2]	Understand water processes - sediments precipitates, etc. Depositional conditions. Chem. environment of soils Organics in sediments, soils.
Geophys./geochem profiles. (lander with rover)	γ-ray spectrometery n-spectrometry EM sounding	Water at surface Water near surface (<1-2 m) Deeper water.
Soil/rock isotopes. (lander with rover)	Mass spectrometry	Age of weathering. Biogenic fractionation.
Definitive trace elements, lithologies, ages, organics, fossils, etc.	Sample return	Environmental conditions, ages, fractionation processes pre-biotic chemistry, past life.

[1] Generic experiment to measure soil oxidents.   [2] Gas Chromatograph Mass Spectrometer

sign instruments to look for organics in lake sediments if you do not know where such sediments are.

In Tables 9-1, 9-2, and 9-3 are listed some aspects of the knowledge needed to further our understanding in the areas of water history, climate change, and the search for life, together with the measurements needed to acquire the knowledge, and why the knowledge would be useful. Most of the instruments listed already exist in some form, although some may be too heavy or consume too much power to be strong immediate candidates. Costs of missions are strongly correlated with spacecraft mass. Landers in the very cost-constrained environments anticipated for the next decade are likely to have payload masses of no more than 20 or 30 kilograms. Fortunately, miniaturization of science instruments is proceeding rapidly. The alpha-proton-X-ray spectrometer, which can analyze all major elements and several minor elements, weighs only 0.6 kg compared with 10 kg for a similar but less capable instrument on the Lunar Surveyor (Economu et al., 1994). Mass spectrometers are also being designed that weigh only 0.6 kg (Young et al., 1994). Although the payload masses will be much reduced from those of a few years ago, instruments are much smaller.

Thus the immediate prospects for Mars exploration are mixed. While funding for a sustained U.S. program seems assured, the funding level is very low compared with previous mission costs. International cooperation is being enthusiastically embraced, but the forging of practical, productive alliances is difficult. Costs will limit greatly the size of future spacecraft and the mass of their payloads, but miniaturization of science instruments and payload substems is proceeding apace. Lastly, while human exploration appears very, very distant, the compulsion to explore will surely carry us to Mars some day.

10 SUMMARY AND CONCLUSIONS

The story of water on Mars impacts almost every facet of the planet's evolution, including accretion, geologic evolution, climatic history, atmospheric dynamics, and biologic potential. It also has major implications for future Mars exploration, both manned and unmanned. There can be little doubt that water has played a major role in the history of Mars, but uncertainties remain about almost every aspect of the water story. There are different possibilities as to how the initial inventory was acquired and how it changed with time. The climatic history of the planet is very uncertain. The stability relations of water in the near-surface materials under present conditions are reasonably well understood, but we do not know the extent to which equilibrium with these conditions has been achieved. The degree to which water is cycled between surface and subsurface reservoirs is unclear. We do not know how much water is near the surface today nor where it is. Despite these uncertainties, some parts of the story of water are understood, and we can place significant constraints on its acquisition and evolution. This chapter summarizes points raised earlier in the book and attempts to highlight both contentious issues and areas of agreement.

Stability Relations

The present thin martian atmosphere and consequent low surface temperatures restrict the amount of water in the atmosphere to an average of close to 10 pr μm and cause the surface to be permanently frozen to depths of roughly 2.5 km at the equator and 6.5 km at the poles. The exact depths depend on thermal conductivities and heat flow. If climatic conditions were the same at the end of heavy bombardment, around 3.8 Gyr ago, the permanently frozen region would have been about one-fifth as thick because of the higher heat flow. Two thirds of the martian surface are

covered with heavily cratered terrain that has survived since the end of heavy bombardment. By analogy with the Moon, these terrains should be underlain by an impact-brecciated megaregolith to a depth of roughly 10 km. The one third of the planet where younger deposits are exposed at the surface should still have a megaregolith at depth. Estimates of the water-holding capacity of the megaregolith range from the equivalent of 0.5 to 1.5 km of water spread evenly over the whole planet, but we do not know how much water is actually present. The only place where water has been unequivocally identified on the surface is at the north pole, where a small, permanent water-ice cap is exposed when the seasonal CO_2 cap disappears in summer. If Mars has or had significant amounts of water, as is likely, then it exists as ice in the shallow megaregolith or as groundwater in the porous megaregolith below the permanently frozen zone.

Under present climatic conditions, ground ice is unstable at latitudes less than 30–40° because mean annual temperatures are above the frost point temperature, which is close to 197 K. The exact latitude depends on the albedo and thermal inertia of the near-surface materials, which vary from place to place. Any ice present within the near-surface materials at these low latitudes will tend to sublime and diffuse into the atmosphere at rates dependent on the permeability of the overlying materials. Imposition of the present climatic conditions must have initiated a dehydration front, which in 4 Gyr would have penetrated to a depth of a few hundred meters. The water that was lost from the ground would have sublimed into the atmosphere and been frozen out at the poles. The layered terrains at the poles are possible repositories for this water. Melting at the base of the polar ice-rich deposits, particularly early in the planet's history when heat flows were higher, could have allowed some of the polar water to re-enter the groundwater system. Over many millions of year this water could have percolated back to the equatorial regions to replace equatorial groundwater lost by incorporation into the cryosphere, thereby establishing a very slow water cycle.

At high latitudes ground ice is permanently stable at depths greater than a few meters. A distinctive terrain softening seen only at high latitudes has been attributed to creep of near-surface materials as a consequence of the presence of ground ice. Absence of similar softening at low latitudes implies ice is absent, as expected from the stability relations. Absence of softening in craters that date from the end of heavy bombardment suggests that ground ice or groundwater was unstable at low latitudes, even at this early period. This does not imply present climate conditions at that time, only that the frost point (or dew point) temperature was exceeded and ground ice (or groundwater) tended to sublime into the atmosphere.

Outflow Channels, Water Inventories, and Deuterium

Outflow channels present the most unequivocal evidence for large quantities of water on the planet. Although many other possible origins have been suggested, the interpretation that they were cut by large floods appears secure. Outflow channels are found in four regions (see Figure 3-1). They have a wide spread of ages, although most formed early in the post–heavy bombardment era. The floods probably had a variety of causes. Some, such as those emerging from the east end of the canyons, may have formed by catastrophic drainage of large lakes. Others start in chaotic terrain, probably as a result of catastrophic eruption of groundwater. Still others, like those northwest of Elysium, appear directly connected in some way with volcanic activity. The peak discharges for the largest floods were very large, ranging from 10^7 to 10^9 m^3 sec^{-1}, compared with roughly 10^5 m^3 sec^{-1} for the peak discharge of the Mississippi. The total volumes of water involved are much more difficult to estimate because we do not know how long the floods lasted nor their erosive efficiency. A very conservative estimate of the amount of water needed to cut the channels around the Chryse basin is 6×10^6 km^3, or the equivalent of 40 m

spread over the whole planet. The actual value is likely to be considerably higher. Best estimates (guesses) for the amount of water required to cut one of the larger channels is n $\times$ 10^5 km^3.

Major controversies with respect to the floods are their timing with respect to each other and the fate of the water. Baker et al. (1991) have suggested that many large floods occurred simultaneously and released such large volumes of water that ocean-sized bodies of water formed in the low-lying northern plains, into which many of the flood channels debouch. They suggested that the oceans formed and dissipated episodically throughout martian history, triggering short-lived but major climatic changes. Support for the hypothesis is found in linear features in the northern plains, interpreted as shorelines; young valley networks, indicative of late warm conditions; and features interpreted as having formed by glaciers. Estimates for the volumes of the oceans range up to 6.5×10^7 km^3, or 450 m spread over the whole planet. There are many reasons to be skeptical of this proposal. The "glacial" and "shoreline" features have other possible interpretations. The young valleys are largely restricted to volcanoes, so that a local cause, such as hydrothermal activity, appears more likely than a global cause, such as climate change. Simultaneity of many large floods seems unlikely, and the mechanisms for climate change, dissipation of the oceans and re-cycling of CO_2 fixed in carbonates are all very obscure. A more conservative scenario is one in which the lakes that formed at the ends of the channels were comparable in volume to individual floods (n $\times$ 10^5 km^3) and froze in place in the low-lying areas. If this is so, then thick ice deposits, representing accumulations from many floods, should be present in the low areas at high latitudes, where most of the floods terminate. In this scenario, once the water has been erupted from below the cryosphere, it is permanently removed from the groundwater system, except for the possible basal melting at the poles mentioned earlier. We should expect, therefore, that the rate of formation of flood features should decrease with time, which is consistent with observations. In this more conservative scenario, the floods are a manifestation of the planet's adjustment to the declining heat flow and climatic conditions imposed at the end of heavy bombardment. The adjustment involved semipermanent transfer of groundwater from the high standing terrains, which are mostly in Tharsis, Elysium, and the southern hemisphere, to the lowest lying areas of the surface, the northern plains and Hellas.

The large floods provide means of estimating the total water inventory at the surface. To the 40 m figure given earlier for the globally averaged amount of water that has passed through the circum-Chryse outflow channels should be added another 10–20 m for the outflow channels elsewhere. This includes only water that actually flowed across the surface in the outflow channels. The actual amount of water present near the surface must be substantially larger than 50–60 m; first, because outflow channels occur only in four regions of the planet, whereas the water is likely to be globally distributed, and second, because the sources of most of the outflow channels are at elevations of 0–1 km above the datum, which leaves most of the global aquifer system at a lower elevation than the flood sources (see Figure 2-8). Most of the water must have remained in the ground because of the lack of hydrostatic head to drive it to the surface. A total inventory of near-surface water of at least a few hundred meters appears reasonable. The contributions to the near-surface inventory from outgassing subsequent to heavy bombardment are likely to be small. Approximately 650×10^6 km^3 of volcanics have been intruded into or extruded onto the crust since the mid-Noachian. If these volcanics contained 1% water, which surely is an upper limit in view of the low water content of the SNCs, and it all outgassed to the atmosphere, which is unlikely, then only an additional 45 m is added. The bulk of the near-surface inventory must have been inherited from the early era of heavy bombardment. If present climatic conditions have been maintained since heavy bombardment, then about 3 m of water have been lost from the upper atmosphere. However, enhanced losses during periods of high obliquity could raise the figure to about 70 m.

Water in the atmosphere is enriched in deuterium because of preferential loss of hydrogen with respect to deuterium from the upper atmosphere. The enrichment by a factor of 5.2 requires that over 90% of the reservoir undergoing enrichment has been lost if the fractionation factor for deuterium loss against hydrogen loss has been the same as at present for most of the planet's history (see Figure 7-10). Water extracted from SNC meteorites has a variable D/H, as though there had been mixing between enriched and nonenriched waters There are two main possible interpretations of the SNC data. The first is that the equivalent of a few hundred meters of water was present at the surface at the end of heavy bombardment, and enrichment of surface waters took place by massive losses of water mostly early in the planet's history. The second possibility is that the surface inventory of water was a few tens of meters at the end of heavy bombardment and that enrichment took place over the course of Mars' entire history. In both cases the variability in D/H in the SNCs results because enriched meteoric water has partly altered primary minerals containing unenriched mantle water.

The SNC deuterium data imply that a large fraction, if not all, of the groundwater is at least partly enriched in deuterium. Even if the episodic ocean hypothesis is correct, the enrichment is unlikely to have been caused by losses during the temporary warm oceanic episodes. For an ocean level close to the 0 km contour and water erupted from a global artesian system, roughly 0.5 km would have been on the ground and 0.5 km would have been left in the ground (see Figure 2-8) for a total near-surface inventory of 1 km. If the first ocean formed of unenriched water and enrichment followed, then several kilometers of water would have had to be present near the surface at the end of heavy bombardment, which is unreasonable because it exceeds the most optimistic estimates of the holding capacity of the megaregolith. If the oceanic hypothesis is correct, the deuterium enrichment must have taken place very early in the planet's history.

If there were at least a few hundred meters of water near the surface at the end of heavy bombardment, as is implied by the geology, and it was all enriched, then the equivalent of at least a few kilometers of water was at the surface at some time between the end of accretion and the end of heavy bombardment, and most of it was lost, thereby enriching what remained. Three factors that could have contributed to such large early losses are hydrodynamic escape, the presence of a thick, water-rich atmosphere, and a high UV output of the young Sun. Any outgassing of mantle water subsequent to the end of heavy bombardment would have diluted the original enrichment, but such contributions are likely to have been small. Similarly, if present conditions are representative of much of Mars' history, then further enrichment of a significant fraction of the near-surface inventory after heavy bombardment, as a result of losses from the upper atmosphere, is likely to have been small because the total hydrogen loss over this time was probably modest compared with the total surface inventory of a few hundred meters. Additions of surface water to the mantle, such as would occur through plate tectonics, have probably also been negligible. Separate histories for magmatic and meteoric water are also implied by the oxygen isotope data.

Accretion, Differentiation, and the Early Atmosphere

The composition of SNC meteorites suggests that Mars accumulated from more volatile-rich materials than the Earth. As the planet grew in size during accretion, impact velocities would have increased until they were so large that volatiles would have been lost from the accreting materials when they impacted the growing planet. After this stage, estimated to have been when the planet had grown to roughly 0.4 times its present radius, some of the volatiles would have been retained at the surface and some would have been incorporated into the planet's interior. In the case of the Earth, a huge (100 bar) steam atmosphere, may have developed from the water released during the impacts. The energy of the impacts, coupled with the blanketing effects of the steam atmosphere, probably caused global melting and a molten surface on the early Earth. Model ages of SNC meteorites show that global differentiation also occurred on Mars at the end of accretion, but whether the surface was molten or not is uncertain. Some models suggest that it was not and that a significant fraction of the devolatilized water was condensed on the surface.

The materials that accumulated to form Mars can be represented by a mix of chondritic meteorites. They were in chemical disequilibrium, including reduced components such as metallic iron and oxidized components such as water. During planetary differentiation and core formation, the interior of the planet would have equilibrated. The metallic iron would have reacted with any water present to form FeO and hydrogen, and the hydrogen would have been driven off into the atmosphere along with other volatiles, such as the noble gases. The interior of the planet may therefore have been left dry and largely depleted of its volatiles, a conclusion that is supported by the low water content of the SNC meteorites. Core formation also resulted in substantial depletions of siderophile and chalcophile elements in the mantle because they mostly entered the core. There is no siderophile excess in the mantle, as appears to be the case with the Earth, so if Mars accreted some small fraction of material late, after core formation, as appears likely if the Earth did, the material did not get folded into the mantle, possibly because of the lack of plate tectonics.

The hydrogen-rich atmosphere formed by equilibration and degassing of the interior was subject to high fluxes of extreme ultraviolet radiation from the young Sun. This resulted in massive thermal escape of hydrogen and loss of other gases by hydrodynamic escape. During the process, most of the noble gases were lost, leaving the remaining highly fractionated. Volatiles, such as H_2O and CO_2, which could either condense on the surface or react with surface materials, may have partly avoided being lost by hydrodynamic escape or by being blown away by large impacts. The rate of hydrodynamic escape declined rapidly with the declining flux of extreme ultraviolet radiation from the Sun and was essentially over by 4.2 Gyr ago. Massive losses of hydrogen during this early era may have caused the remaining surface waters to be partly enriched in deuterium.

Conditions in the middle of the era of heavy bombardment are very unclear. The planet had to emerge from heavy bombardment with at least a few hundred meters of water at the surface. If climates were such that liquid water was present, high rates of weathering would have occurred and CO_2 would have reacted with surface materials to form carbonates. The carbonates and weathered materials would have tended to be folded deep into the megaregolith by impacts. Thick, easily eroded deposits in various places on the surface at present and the almost omnipresent drift material may be largely the products of weathering that occurred at the end of this era.

High rates of volcanism are also likely at this time, and this may have added to the volatile inventory at the surface, although most of the outgassing should have taken place during the early catastrophic outgassing that accompanied core formation. High rates of erosion and valley formation indicate that atmospheric conditions during this era were very different from those that prevailed for most of the time after the end of heavy bombardment. Hydrogen losses from a thick, water-rich atmosphere during late heavy bombardment could have further contributed to the deuterium enrichment.

Valley Networks, Climate Change, and Carbonates

Long-term climate changes are supported by two main pieces of evidence, the presence of valley networks and high erosion rates early in the planet's history. If the valley networks are fluvial, that is, they were cut by streams and rivers, then warmer climates almost certainly are required for their formation. Although it has been suggested that rivers could cut the valleys under present climatic conditions if the rivers were spring fed, this is contrary to terrestrial experience of spring-fed streams under subzero temperature conditions and fails to explain how the groundwater system remained charged with water. The rate of valley formation declined rapidly at the end of heavy bombardment, but valleys continued to form occasionally, later in the planet's history, particularly on steep slopes, such as crater and canyon walls, and on some volcanoes. Even in the oldest, most heavily dissected regions, drainage densities are two to three orders of magnitude less than are typical for the Earth. The valley networks are smaller than terrestrial drainage basins. For over 90% of the networks, the longest path through the network is under 200 km.

The consensus view is that the networks are fluvial. In Chapter 4, an alternative view, that mass wasting has played a major role in valley network formation, was discussed. This personal view is supported by the lack of river channels within the valleys, the rectangular or U-shaped cross section of the valleys, the presence in some valleys of longitudinal ridges and levee-like features, and the lack of small (<1 km wide) tributaries in most upland valleys. What is envisaged is a process of headward erosion by mass wasting, aided by the movement of groundwater, and slow down-valley transport of the debris, also by mass wasting, and also aided by the presence of groundwater. Groundwater is essential to the process, and water may have occasionally flooded to the surface, as in the outflow channels, and helped to carve the valleys.

Erosion rates since the end of heavy bombardment have typically been extremely low, no more than 10^{-2} μm yr^{-1}, which is three to six orders of magnitude less than terrestrial rates. Craters that formed after the end of heavy bombardment are almost all perfectly preserved, retaining fine depositional details of their ejecta blankets. In contrast, high erosion rates prevailed during heavy bombardment. Even large craters that formed at this time show a wide range of degradation, some being barely discernible. Although erosion rates for this era are difficult to quantify, because of uncertainties in the cratering rate, the best estimate is that rates were about 10 μm yr^{-1}, near the low end of the terrestrial range. This major change in erosion rates at the end of heavy bombardment is confirmed by crater counts. Although the total number of craters on the ancient terrains that formed during heavy bombardment is large, the number of craters that preserve details of their ejecta is roughly the same as the total number of craters on surfaces that formed immediately after heavy bombardment. There can be little doubt that an enormous change in the rate of landscape modification took place at the end of heavy bombardment.

The high rates of valley formation and high erosion rates at the end of heavy bombardment have both been attributed to warm, wet climatic conditions at that time. The prevailing view has been that Mars had a thick CO_2–H_2O atmosphere that caused surface temperatures to be above freezing, and that the atmosphere collapsed soon after the end of heavy bombardment, as a result of fixation of CO_2 in carbonates. Prior to the end of heavy bombardment the atmosphere was sustained by the rapid return of CO_2 fixed in carbonates to the atmosphere because of high rates of burial of the carbonates by volcanics and impact ejecta, and the steep thermal gradient that prevailed at that time. This view is being challenged on two grounds. First, a thick atmosphere, early in the planet's history when impact rates were high, would tend to be lost as a result of impact erosion. Second, recent modeling suggests that on early Mars a CO_2–H_2O atmosphere could not raise surface temperatures even close to freezing. With the lower solar constant expected for that time, cloud formation would cause an increase in the planetary albedo, which would offset greenhouse effects, and surface temperatures would remain cold no matter how thick the atmosphere. The lifetimes in the atmosphere of other potential greenhouse gases, such as NH_3, CH_4, and SO_2, are too short to maintain warm surface temperatures. Either early Mars was not warm and wet, or some significant factor has not been accounted for in the modeling.

The story on carbonates is equivocal. Spectral data indicate that the airborne dust contains 1–3% carbonate and that no more than 3% of the surface materials can be calcium carbonate. Other spectral data have been interpreted as demonstrating the presence of 10–20% hydrous magnesium carbonate in the surface materials. All the SNC meteorites contain small quantities of carbonate, which suggests that the groundwater is charged with CO_2. Some geologic features, particularly in the canyons, have been interpreted as formed by solution of carbonates.

Interpretation of the valley networks that significantly postdate the end of heavy bombardment presents another problem with respect to climate. If they are fluvial and cut by slow erosion of running water, then major climatic excursions, such as those proposed under the episodic ocean hypothesis, are probably required. Valleys that are younger than lower Hesperian constitute significantly less than 10% of all the valleys, and most are on steep slopes or volcanoes. There are three main alternative explanations to the fluvial hypothesis with major climatic excursions. The first is that the young valleys formed from spring-fed streams under present climatic conditions. I consider this unlikely for the reasons given earlier. The second is that some of the valleys, particularly those on volcanoes, formed by spring-fed streams of water that had been warmed by hydrothermal activity, are not so susceptible to freezing. The third alternative is that the valleys formed in easily erodible deposits, such as ash, mainly by mass wasting, assisted by the presence of groundwater.

Quasi-periodic changes in the orbital and rotational motions of the planet must have induced atmospheric changes throughout the history of the planet as CO_2 was redistributed between its various reservoirs. The obliquity is thought to vary chaotically, occasionally reaching values as high as 60° and as low as 0°. At low obliquities, volatiles such as CO_2 and H_2O will tend to accumulate near the poles; at high obliquities they will tend to be driven from the poles. Only modest amounts of CO_2 are likely to be present in the three known reservoirs: the atmosphere, the cap, and the regolith. If these are the only significant sources of CO_2, obliquity changes are unlikely to increase atmospheric pressure by more than a few tens of millibars, which would have a negligible effect on surface temperatures. The polar layered terrains could, however, contain large amounts of clathrate, although there is no supporting evidence for this. If true, then at high obliquities the layered terrains would dissipate and surface pressures could rise to a few tenths of a bar.

Lakes, Hydrothermal Systems, and Life

The prospects that life could have started on Mars under present conditions are small. The present surface conditions are probably representative of much of the planet's history, and they are very hostile to life because of the high UV flux, the oxidizing nature of the soil, and the low temperatures, which make liquid water unavailable. The prospects for life in the distant past are, however, much better. High erosion rates and high rates of valley formation suggest that climatic conditions may have been different at the end of and immediately after heavy bombardment, when life started here on Earth. Warm, wet subsurface conditions were probably also common because of high rates of volcanism and abundant groundwater. Warm, chemically rich surface springs may also have been common. Thus, many of the environmental niches present on Earth when life started were also present on early Mars.

Studies of molecular phylogeny suggest that the life forms on Earth that have evolved least from the common primitive ancestor of all terrestrial life are anaerobic hypothermophiles that grow in high-temperature environments, such as hot springs, fixing CO_2 and utilizing geochemical sources such as sulfur and hydrogen. Given that the surface of Mars is water rich and that high rates of volcanism must have prevailed early in the planet's history because of the high heat flow, hydrothermal activity must have been common. Hydrothermal activity provides two broad and varied enviroments hospitable to life: the warm and wet subsurface, and warm surface springs. Hydrothermal activity would have declined with time along with the heat flow. Various geologic features have been attributed to volcano-ice interactions. Valleys on young volcanoes, for example, may have resulted from seepage onto the surface of warm, volcanically heated waters. But the detection of past hydrothermal activity requires detection of hydrothermal minerals, and the appropriate observations have yet to be made. Some alteration products in SNC meteorites probably are the result of interactions with warm, meteoric water, as indicated by their mineralogy and deuterium content.

Lakes are another environment in which life may have started or survived. Three types of lakes have been identified: those that likely formed in areas of convergent drainage by valley networks in the heavily cratered uplands, those that must have formed at the end of large outflow channels, and those that formed within the canyons. The lakes suggested by convergent drainage in the cratered uplands are of particular biologic interest because they are ancient and formed at a time when climatic conditions were possibly more benign than at present. The lakes that formed at the end of the large outflow channels are probably of minimal biologic interest. They are considerably younger than the postulated lakes in the cratered uplands. If climatic conditions were the same as at present when they formed, they would have rapidly iced over and then slowly frozen, and life is unlikely to have had time to start. Even with the episodic ocean hypothesis, the oceans that are postulated are thought to have been short lived. The lakes within the canyons present a more likely habitat than those at the end of outflow channels, although they also formed well after the end of heavy bombardment, the time for which we have the best evidence for different climates. Their former presence is indicated by thick stacks of sediment within the canyons. The lakes could have survived under an ice cover for long periods, possibly hundreds of millions of years, even under present climatic conditions, because seepage from groundwater would have replaced water lost from the surface by ablation.

If life started earlier, when climatic and volcanic conditions were more favorable, it could have survived within the canyon lakes long after favorable environments had disappeared elsewhere.

The future

Our success in answering many of the questions raised in this book will depend largely on future Mars missions. Planetary missions have goals other than pure science. They may, for example, be used to support foreign policy, to promote technology development, or to stimulate education and interest in science and technology. For Mars, another goal is to prepare the way for eventual human exploration. The science goals of Mars exploration can be grouped into three main themes: to determine the mode of formation and evolution of the solid planet, to determine the planet's climatic history, and to determine its biologic history, if any. These goals can best be addressed by an exploration program that concentrates first on completing the global remote sensing from orbit, then on deploying simple landers to the surface to make pilot measurements at crucial locations, and then to undertake more complex landed missions, such as long-range rovers, networked landers, and sample return. The mix of missions will depend on what resources are available and what goals are emphasized.

To resolve issues with respect to water, we clearly want to get closer looks at water-formed features, such as valley networks and lake sediments, and at regions such as the low-lying northern plains where water has pooled in the past. We need to locate and observe past and present hydrothermal deposits. We want to detect where water is now. Is water pervasive in the high-latitude regolith at shallow depths, as suggested by terrain softening? Are there thick ice deposits at the end of outflow channels? Is water-ice a major component of the polar layered terrains? Inventories of other volatiles, such as carbonates and nitrates, are also of interest for climate history. Are there thick carbonate deposits and if so where are they and what are their ages? Samples of the former atmospheres may be preserved in different-aged rocks. Water lain sediments, evaporites, and hydrothermal deposits will all give indications of past conditions in water-rich environments. Past climatic conditions may also be decipherable from the mineralogy of weathered products. Multiple weathering horizons in soils would indicate past climatic variations. We therefore need to make both close in situ observations and detailed comprehensive analyses, including age determinations, of a wide variety of rocks and soils. Direct evidence of the critical early events of accretion and heavy bombardment will have been destroyed by subsequent events, but a record is still preserved in the chemistry and isotopics of the atmosphere and surface materials. The isotopics also preserve a record of losses from the upper atmosphere. Biologic interest is mainly in past lacustrine and hydrothermal deposits, so these should be examined closely for evidence of chemical disequilibria, prebiotic chemistry, and fossil life. Should evidence of past life be found, the emphasis might shift to looking for present-day life in favorable niches, such as warm, wet volcanic vents.

At the time of writing, the United States had initiated a long-term, but very cost-constrained, Mars exploration program. Russia, Japan, and the European Space Agency also had plans for missions to Mars. The global remote sensing was expected to have been largely completed by the end of the century, after which time the emphasis is expected to be mainly on surface exploration. While enthusiasm for Mars exploration is high, resources are limited. To mount missions with complex capabilities, such as long-range rovers and sample return, will present major challenges. Meanwhile, human exploration appears very distant but not forgotten.

Thus, understanding the story of water is crucial for the three main concerns of Mars science: the formation and evolution of the solid planet, elucidation of its climate history, and the search for life. From the time people started to speculate about what Mars is like, water has been the central issue. While our understanding has grown enormously since the first spacecraft visited the planet over 30 years ago, the major questions have changed little. While 18th and 19th century philosphers speculated about maria, canals, weather, and advanced civilizations, we argue about oceans, channels, climate, and the origin of life. The unique fascination of Mars remains. Mars still beckons* and surely sometime in the 21st century we will heed its call and see not only machines but people on the Chryse plains.

*Mars Beckons, Knopf, 1990. A book by the New York Times science correspondent John Noble Wilford on the allure of Mars.

REFERENCES

Abe, Y., and Matsui, T., 1985, The formation of an impact-generated H_2O atmosphere and its implications for the early thermal history of the Earth. J. Geophys. Res., 90, C545–C559.

Abe, Y., and Matsui, T., 1986, Early evolution of the Earth: Accretion atmospher formation and thermal history. J. Geophys. Res., 91, E291–E302.

Abrahams, A. D., 1984, Channel networks: A geomorphological perspective. Water Resources Res., 20, 161–168.

Achenback-Richter, L., Gupta, R., Steter, K., and Woese, C., 1987, Were the original Eubacteria thermophiles?, Syst. Appl. Microbiol., 9, 34–39.

Ahrens, T. J., O'Keefe, J. D., and Lange, M. A., 1989, Formation of atmospheres during accretion of the terrestrial planets. in Origin and Evolution of Planetary and Satellite Atmospheres (Atreya, S. K., et al., eds), University of Arizona Press, Tucson, pp. 328–385.

Allen, C. C., 1979 Volcano-ice interactions on Mars. J. Geophys. Res., 84, 8048–8059.

Anders, E., and Ebihara, M., 1982, Solar system abundances of the elements. Geochim, Cosmochim. Acta, 46, 2363–2380.

Anders, E., and Grevesse, N., 1989, Abundances of the elements. Meteoritic and solar. Geochim. Cosmochim. Acta, 53, 197–214.

Anders, E., and Owen, T., 1977, Mars and Earth: Origin and abundance of volatiles. Science, 198, 453–465.

Anderson, D. M., Gaffney, E. S., and Low, P. F., 1967, Frost phenomena on Mars. Science, 155, 314–322.

Anderson, D. M., and Tice, A. R., 1973, The unfrozen interfacial phase in frozen soil water systems. In Ecological Studies Anaysis and Synthesis, Vol 4 (Hadasa A., et al., eds), Springer-Verlag, New York, pp. 107–124.

Arvidson, R. E., Guiness, E. A., and Lee, S., 1979, Differential eolian redistribution rates on Mars. Nature, 278, 533–535.

Bagnold, R. A., 1966, An approach to the sediment transport problem from general physics. U. S. Geol. Survey, Prof. Paper 422–1.

Baker, V. R., 1973, Paleohydrology and sedimentology of Lake Missouls flooding of western Wahington. Geol. Soc. Am. Spec. Paper 144.

Baker, V. R., 1982, The channels of Mars. University of Texas Press, Austin.

Baker, V. R., 1990, Spring sapping and valley network development. Geol. Soc. Am. Spec. Paper 252, pp. 235–265.

Baker, V. R., and Milton, D. J., 1974, Erosion by catastrophic floods on Mars and Earth. Icarus, 23, 27–41

Baker, V. R., and Partridge, J. B., 1986, Small martian valleys: Pristine and degraded morphology. J. Geophys. Res., 91, 3561–3572

Baker, V. R., Strom, R. G., Gulick, V. C., Kargel, J. S., Komatsu, G., and Kale, V. S., 1991, Ancient oceans, ice sheets and the hydrological cycle on Mars. Nature, 352, 589–594.

Baker, V. R., Carr, M. H., Gulick, V. C., Williams, C. R., and Marley, M. S., 1992, Channels and valley networks. In Mars (Kieffer, H. H., et al., eds.), University of Arizona Press, Tucson, pp 493–522.

Baker, V. R., Benito, G., and Rudoy, A. N., 1993, Paleohydrology of Late Pleistocene superflooding, Altay Mountains, Siberia. Science, 259, 348–350.

Banin, A., Clark, B. C., and Wänke, H., l992, Surface chemistry and mineralogy. In Mars (Kieffer, H. H., et al., eds.), University of Arizona Press, Tucson, pp. 594–625.

Barlow, N. G., 1988a, Crater size-frequency distributions and a revised martian relative chronology. Icarus, 75, 285–305.

Barlow, N. G., 1988b, Parameter affecting formation of martian crater ejecta morphology. Lunar Planet. Sci. Cont. XIX, pp. 31–32.

Barker, E. S., Schorn, R. A., Woszczyk, A., Tull, R. G., and Little, S. J., 1970, Mars: Detection of water vapor during the southern hemisphere spring and summer season. Science, 170, 1308–1310.

Becker, R. H., and Pepin, R. O., 1984, The case for a martian origin of the Shergottites: Nitrogen and noble gases in EETA79001. Earth Planet. Sci. Lett., 69, 225–242.

Becker, R. H., and Pepin, R. O., 1986, Nitrogen and light noble gases in Shergotty. Geochim. Cosmochim. Acta, 50, 993–1000.

Benlow, A., and Meadows, A. J., 1977, The formation of the atmospheres of the terrestrial planets by impacts. Astrophys. Space Sci., 46, 293–300.

Biemann, K., Oro, J., Toulmin, P., Orgel, L. E., Nier, A.O., Anderson, D. M., Simmonds, P. G., Flory, D., Diaz, A. V., Rushneck, D. R., Biller, J. E., and Lafleur, A. L., 1977, The search for organic substances and inorganic volatile compounds in the surface of Mars. J. Geophys. Res., 82, 4641–4658.

Binder, A. B., and Lange, M. A., 1980, On the thermal history, thermal state, and related tectonism of a moon of fission origin. J. Geophys. Res., 85, 3194–3208.

Binder, A. B., Arvidson, R. E., Guiness, E. A., Jones, K. L., Morris, E. C., Pieri, D. C., and Sagan, C., 1977, The geology of the Viking 1 landing site. J. Geophys. Res., 82, 4439–4451.

Bjoraker, G. L., Mumma, M. J., and Larson, H. P., 1989, The value of D/H in the Martyian atmosphere: Measurements of HDO and H_2O using the Kuiper Airborne Observatory. Proc. 4th Int. Conf. on Mars., Tucson, Jan, 10-13, 1989, pp. 69–70.

Blaney, D. L., and McCord, T. B., 1989, An observational search for carbonates on Mars. J. Geophys. Res., 94, 10159–10166.

Blasius, K. R., and Cutts, J. A., 1979, Erosion and transport in martian outflow channels. Lunar Planet. Sci. Conf. X, pp. 140–141.

Blasius, K. R., and Cutts, J. A., 1980, Global patterns of primary crater ejecta morphology on Mars. NASA Tech. Memo. 82385, pp. 147–149.

Blasius, K. R., Cutts, J. A., Guest, J. E., and Masursky, H., 1977, Geology of the Valles Marineris: First analysis of imaging from the Viking 1 orbiter primary mission. J. Geophys. Res., 82, 4067–4091.

Bogard, D. D., and Johnson, P., 1983, Martian gases in an Antarctic meteorite. Science, 221, 651–654.

Bogard, D. D., Nyquist, L. E., and Johnson, P., 1984, Noble gas content of Shergottites and implications for the martian origin of SNC meteorites. Geochim. Cosmochim. Acta, 48, 1723–1739.

Boothroyd, A. I., Sackmann, I.-J., and Fowler, W. A., 1991, Our Sun II. Early mass loss of 0.1 M_0 and the case of hte missing lithium. Astrophys. J., 377, 318–329.

Boston, P. J., Ivanov, M. V., and McKay, C. P., 1992, On the possibility of chemosynthetic ecosystems in subsurface habitats on Mars, Icarus, 95, 300–308.

Brackenridge, G. R., 1990, The origin of fluival valleys and early geologic history, Aeolis Quadrangle, Mars. J. Geophys. Res., 95, 17, 289–317, 308.

Brackenridge, G. R., 1993, Ancient martian valley genesis and paleoclimatic inference: The present as a key to the past. Lunar Planet Inst. Tech Rept. 93–03, Pt 1, p. 2.

Brackenridge, G. R., Newsom, H. e., and Baker, V. R., 1985, Ancient hot springs on Mars: Origins and paleosignificance of small martian valleys. Geology, 13, 859–862.

Brass, G. W., 1980, Stability of brines on Mars. Icarus, 42, 20–80.

Brett, R., 1984, Chemical equilibration of the Earth's core and upper mantl;e. Geochim. Gimochim Acta, 48, 1183–1188.

Brunsden, D., and D. B. Prior, 1984, Slope Instability. John Wiley, Chichester.

Burns, R. G., 1988, Gossans on Mars. Proc. Lunar Planet. Sci. Conf., 18, 713–721.

Cairns-Smith, A. G., 1982, Genetic Takeover and the Mineral Origins of life. Cambridge University Press, Cambridge.

Calvin, W. M., King, T. V. V., and Clark, R. N., 1994, Hydrous carbonates on Mars?: Evidence from Mariner 6/7 infrared spectrometer and ground-based telescopic spectra. J. Geophys. Res., 99, 14659–14675.

Carr, M. H., 1973, Volcanism on Mars. J. Geophys. Res., 78, 4049–4062.

Carr, M. H., 1974, The role of lava erosion in the formation of lunar rilles and martian channels. Icarus, 22, 1–23.

Carr, M. H., 1979, Formation of Martian flood features by release of water from confined aquifers. J. Geophys. Res., v. 84, p. 2995–3007.

Carr, M. H., 1981, The Surface of Mars. Yale University Press, New Haven.

Carr, M. H., 1983, The stability of streams and lakes on Mars. Icarus, 56, 476–495.

Carr, M. H., 1986, Mars: A water rich planet? Icarus, v. 56, 187–216.

Carr, M. H., 1987, Water on mars. Nature, v. 326, 30–35.

Carr, M. H., 1989, Recharge of the early atmosphere of Mars by impact-induced release of CO_2. Icarus, 79, 311–327.

Carr, M. H., 1990, D/H on Mars: Effects of floods, volcanism, impacts and polar processes. Icarus, 87, 210–227

Carr, M. H., 1992, Post Noachian erosion rates: Implication for Mars climate change. (Lunar Planet. Sci. Conf. XXIII, pp. 205–206.

Carr, M. H., 1995, The martian drainage system and the origin of networks and fretted channels. J. Geophys. Res. (in press)

Carr, M. H., and Clow, G. D., 1981, Martian channels and valleys: Their characteristics, distribution and age. Icarus, 48, 91–117.

Carr, M. H., and Greeley, R., 1980, Volcanic features of Hawaii. NASA SP-403.

Carr, M. H., and Schaber, G. G., 1977, Martian permafrost features. J. Geophys. Res., 82, 4039–4054.

Carr, M. H., and Wänke, H., 1992, Earth and Mars: Water inventories as clues to accretional histories. Icarus, 98, 61–71.

Carr, M. H., Crumpler, L., Cutts, J. A., Greeley, R., Guest, G. E., and Masursky, H., 1977a, Martian impact craters and implacement of ejecta by surface flow. J. Geophys. Res., 82, 4055–4065.

Carr, M. H., Greeley, R., Blasius, K. R., Guest, J. E., and Murray, J. B., 1977b, Some martian volcanic features as viewed from the Viking orbiters. J. Geophys. Res., 82, 3985–4015.

Carr, M. H., Wu, S. C., Jordan, R., and Schafer, F. J., 1987, Volumes of channels, canyons and chaos in the circum-Chryse region of Mars. Lunar Planet. Sci. conf. XVIII, pp. 155–156.

Cess, R. D., Ramanathan, V., and Owen, T., 1980., The martian paleoclimate and enhanced carbon dioxide. Icarus, 41, 159–165.

Chang, S., 1982, Priobiotic organic matter: Possible pathways for synthesis in a geological context. Phys, Earth Planet. Int., 29, 261–280.

Chang, S., 1988, Planetary environments and the conditions of life. Phil. Soc. Trans. R. Soc. Lond, A 325, 601–610.

Chang, S., DesMarais, D., Mack, R., Miller, S. L., and Strathearn, G. E., 1983, Prebiotic organic syntheses and the origin of life. In The Earth's Earliest Biosphere (Schopf, J. W., ed.), Princeton University Press, Princeton, NJ, pp. 53–92.

Chen J. H., and Wasserburg, G. J., 1986, Formation ages and evolution of Shergotty and its parent planet from U-Th-Pb systematics. Geochim. Cosmochim. Acta, 50, 955–968.

Chyba, C. F., 1990, Impact delivery and erosion of planetary oceans in the early inner solar system. Nature, 343, 129–133.

Chyba, C. F., 1991, Terrestrial mantle siderophiles and the lunar impact record. Icarus, 92, 217–233.

Chyba, C. F., and Sagan, C., 1992, Endogenous production, exogenous delivery and impact-shock synthesis of organic molecules: An inventory for the origins of life. Science, 355, 125–132.

Clark, B. C., and Van Hart, D. C., 1981, The salts of Mars. Icarus, 45, 370–378.

Clayton, R. N., and Mayeda, T. K., 1983, Oxygen istotopes in euchrites, shergottites, nakhlites and chassignites. Earth Planet. Sci. Lett., 62, 1–6.

Clow, G. D., 1987, Generation of liquid water on Mars through the melting of a dusty snowpack. Icarus, 72, 95–127.

Clifford, S. M., 1981, A pore volume estimate of the martian megaregolith based on a lunar analog. Lunar Planet. Sci. Inst. Contrib. 441, 46–48.

Clifford, S. M., 1982, Mechanisms for the vertical transport of H_2O in the martian regolith. Lunar Planet. Sci. Inst. Contrib. 488, 23–24.

Clifford, S. M., 1987, Polar basal melting on Mars. J. Geophys. Res., 92, 9135–9152.

Clifford, S. M., 1993, A model for the hydrologic and climatic behavior of water on Mars. J. Geophys. Res., 98, 10,973–11,016.

Clifford, S. M., and Fanale, F. P., 1985, The thermal conductivity of the martian crust. Lunar Planet. Sci. Conf. XVI, pp. 144–145.

Clifford, S. M., and Hillel, D., 1983, The stability of ground ice in the equatorial regions of Mars. J. Geophys. Res., 88, 2456–2474.

Clifford, S. M., and Zimbelman, J. R., 1988, Softened terrain on Mars. Lunar Planet. Sci. Conf. XIX, pp. 199–200.

Cloud, P., 1968, Atmospheric and hydrospheric evolution on the primitive earth. Science, 160, 729–736.

Cloud, P., 1983, Early biogeologic history: Emergence of a paradigm. In The Earth's Earliest Biosphere (Schopf, J. W., eds), Princeton University Press, Princeton, NJ, pp. 14–31.

Collins, S. A., 1971, The Mariner 6 and 7 pictures of Mars. NASA SP-263.

COMPLEX, 1978, Strategy for Exploration of the Inner Planets: 1977–1978. National Academy of Sciences, Washington, D.C.

COMPLEX, 1990, 1990 Update to Strategy for Exploration of the Inner Planets. National Academy of Sciences, Washington, D.C.

Costard, F. M., 1988, Thickness of sedimentary deposits of the mouth of outflow channels. Lunar Planet. Sci. Conf. XIX, pp. 211–212.

Costard, F. M., and Kargel, J. S., 1995, Outwash plains and thermokarst on Mars. Icarus, 114, 93–112.

Craddock, R. A., and Maxwell, T. A., 1993, Geomorphic evolution of the martian highlands through ancient fluvial processes. J. Geophys. Res., 98, 3453–3468.

Cronin, J. R., 1989, Origin of organic compounds in carbonaceous chondrites. Adv. Space Res., 9, (2)59–(2)64.

Cutts, J. A., 1973, Nature and origin of layered deposits in the martian polar regions. J. Geophys. Res., 78, 4231–4249.

Cutts, J. A., and Blasius, K. R., 1979, Martian outflow channels: Quantitative comparison of erosive capacities for eolian and fluvial models. Lunar Planet. Sci. Conf. X, pp. 257–259.

Cutts, J. A., and Lewis, B. H., 1982, Models of climatic cycles record in Martian polar layered deposits. Icarus, 50, 216–244.

Dacey, M. F., and Krumbein, W. C., 1976, Three growth models for stream channel networks. J. Geol., 84, 153–164.

Davies, D. W., 1981, The Mars water cycle. Icarus, 45, 398–414.

Davis, P. A. and L. A. Soderblom, 1984. Modeling crater topography and albedo from monoscopic Viking orbiter images. 1. Methodology. J. Geophys. Res., 89, 9449–9457.

Dreibus, G., and Wänke, H., 1987, Volatiles on Earth and Mars: A comparison. Icarus, 71, 225–240.

Dunne, T., 1990, Hydrology, mechanics, and geomorphic implications of erosion by subsurface flow. Geol. Soc. Am. Spec. Paper 252, pp. 1–28

Dzurisin, D., and Blasius, K. R., 1975, Topography of the polar layered deposits of Mars. J. Geophys. Res., 80, 326–3306.

Economu, T., Turkevich, A., Rieder, R. and Wänke, H., 1994, Chemical composition of the martian surface: The APX-Spectrometer. Jet Propulsion Lab. Tech. Rep. D-12017, pp. 45–46.

Elder, J. W., 1965, Physical Processes in Geothermal Areas. American Geophysical Union, Monograph 8, pp. 211–239.

Ellis, A. J., and Mahon, W. A. J., 1977, Chemistry and Geothermal Systems. Academic Press, New York.

Environmental Systems Research, 1992, Understanding GIS. Environmental Systems Research, Redlands, CA, 536 p.

Ernst, W. G., 1983, The early Earth and the Archean rock record. in The Earth's Earliest Biosphere (Schopf, J. W., ed), Princeton University Press, Princeton, NJ, pp. 41–52.

Ezell, E. C., and Ezell, L. N., 1984, On Mars. NASA SP-4212, 535 p.

Fanale, F. P., 1976, Martian volatiles: Their degassing history and geochemical fate. Icarus, 28, 179–202.

Fanale, F. P., and Cannon, W. A., 1979, Mars: CO_2 adsorption and capillary condensation on clays: Significance for volatile storage and atmospheric history. J. Geophys. Res., 83, 2321–2325.

Fanale, F. P., Salvail, J. R., Banerdt, W. B., and Saunders, R. S., 1982, The regolith-atmosphere-cap system and climate change. Icarus, 50, 381–407.

Fanale, F. P., J. R. Salvail, A. P. Zent, and S. E. Postawko, 1986, Global distribution and migration of sub-surface ice on Mars. Icarus, 67, 1–18

Fanale, F. P., Postawko, S. E., Pollack, J. B, Carr, M. H., and Pepin, R. O., 1992, Mars: Epochal climate change and volatile history. In Mars (Kieffer, H. H., et al., eds), University of Arizona Press, Tucson, AZ, pp. 1135–1179.

Farmer, C. B., and Doms, P. E., 1979, Global and seasonal water vapor on Mars and implications for permafrost. J. Geophys. Res., 84, 2881–2888.

Farmer, C. B., Davies, D. W., and LaPorte, D. D., 1976, Mars: Northern summer ice cap—water vapor observations from Viking 2. Science, 194, 1399–1341.

Feigelson, E. D., and Kriss, G. A., 1989, Soft X-ray observations of pre-main sequence stars in the Chamaeleon dark cloud. Astron. J., 338, 262–276.

Flammarion, C., 1892, La Planéte Mars et ses Conditions d'Habitabilités, Vol. 1. Gauthier-Villars et Fils, Paris.

Fox, J. L., 1993, The production and escape of nitrogen atoms on Mars. J. Geophys. Res., 98, 3297–3310.

Fox, J. L., and Dalgarno, A., 1983, Nitrogen escape from Mars. J. Geophys. Res., 88, 9027–9032.

French, H. M., 1976, The Periglacial Environment. Longman, London.

Friedmann, E. I., 1982, Endolithic microorganisms in the Antarctic cold desert. Science, 215, 1045-1053.

Fuller, 1922, Some unusual erosion features in the loess of China. Geogr. Rev., 12, 570–584.

Ganapathy, R., and Anders, A., 1974, Bulk composition of the Moon and Earth, estimated from meteorites. Proc. 5th Lunar Sci. Conf., pp. 1181–1206.

Gault, D. E., and Greeley, R., 1978, Exploratory experiments of impact craters formed in viscous-liquid targets: Analogs for martian rampart craters? Icarus, 34, 486–495.

Gerasimov, M. V., and Mukhin, L. M., 1979, On the mechanism of atmosphere formation during the accretion of the Earth and the terrestrial planets. Sov. Astron. Lett. 5, 2212–2223.

Gierasch, P. J., 1974, Martian dust storms. Rev. Geophys. Space Phys., 12, 730–734.

Goettel, K. A., 1981, Density of the mantle of Mars. Geophys. Res. Lett., 8, 497–500.

Gogarten, J. P., Kibak, H., Dittrich, P., Taiz, L., Bowman, E. J., Bowman, B. J., Manolson, M. F., Poole, R. J., Date, T., Oshima, T., Konish, J., Denda, K., and Yoshida, M., 1989, Evolution of the vacuolar H^+-ATPase: Implications for the origin of eukaryotes. Proc. Natl. Acad. Sci. USA, 86, 6661–6665.

Goldspiel, J. M., and Squyres, S. W., 1991, Ancient aqueous sedimentation on Mars. Icarus, 89, 392–410.

Goldspiel, J. M., Squyres, S.W., and Jankowski, D. G., 1993, Topography of small martian valleys. Icarus, 105, 479–500.

Gooding, J. L., 1992, Soil mineralogy and chemistry on Mars: Possible clues from salts and clays in SNC meteorites. Icarus, 99, 28–41.

Gooding, J. L., Wentworth, S. J., and Zolensky, M. E., 1988, Calcium carbonate and sulfate of possible extraterrestrial origin in EETA79001 meteorite. Geochim. Cosmochim. Acta, 52, 909–915.

Gooding, J. L., Wentworth, S. J., and Zolensky, M. E., 1991, Aqueous alteration of the Nakhla meteorite. Meteoritics, 26, 135–143.

Gooding, J. L., Arvidson, R. E., and Zolotov, M. Y., 1992, Physical and chemical weathering. In Mars (Kieffer, H. H., et al., eds), University of Arizona Press, Tucson, pp. 626–651.

Gough, D. O., 1981, Solar interior structure and liminosity variations. Solar Phys., 74, 21–34.

Graf, W. H., 1971, Hydraulics of sediment transport. McGraw-Hill, New York.

Grant, J. A., and Schultz, P. H., 1993, Degradation of selected terrestrial and martian impact craters. J. Geophys. Res., 98, 11025–11042.

Greeley, R., 1987, Release of juvenile water on Mars: Estimated amounts and timing associated with volcanism. Science, 236, 1653–1654.

Greeley, R., and Crown, D. A., 1990, Volcanic geology of Tyrrhena Patera, Mars. J. Geophys. Res., 95, 7133–7149.

Greeley R., and Guest, J. E., 1987, Geologic map of the eastern equatorial region of Mars. U.S. Geological Survey, Misc Inv. Map I-1802-B

Greeley, R., and Schneid, B. D., 1991, Magma generation on Mars: Amountsss, rates and comparisons with Earth, Moon and Venus. Science, 254, 996–998.

Greeley, R., Lancaster, N., Lee, S., and Thomas, P., 1992, Martian aeolian processes, sediments and features. In Mars (Kieffer, H. H., et al., eds), University of Arizona Press, Tucson, pp. 730–766.

Gregory, K. J., 1976, Drainage networks and climate. In Geomorphology and Climate (Derbyshire, E., ed), Wiley, New York, pp. 289–315.

Gregory, K. J., and Gardiner, V., 1975, Drainage density and climate. Z. Geomorph., 19, 287–298.

Griffith, L. L., Arvidson, R. E., and Shock, E. L., 1995, Hydrothermal carbonate on Mars (in press).

Grossman, J. N., 1988, Formation of chondrules. In Meteorites and the Early Solar System. (Kerridge, J. F., and Matthews, M. S. eds). University of Arizona Press, Tucson, pp. 680–696.

Gulick, V. C., 1992, Magmatic intrusions and hydrothermal systems on Mars. Lunar Planet. Inst. Tech. Rept. 92-02, pp. 50–51

Gulick, V. C., and Baker, V. R., 1989, The role of hydrothermal circulation in the formation of fluvial valleys on Mars. Lunar Planet. Sci. Conf. XX, pp. 369–370

Gulick, V. C., and Baker, V. R., 1990, Origin and evolution of valleys on martian volcanoes. J. Geophys. Res., 95, 14,325–14,344

Gulick, V. C., and Baker, V. R., 1992, Martian hydrothermal systems, some physical considerations. Lunar Planet. Sci. Cont. XXIII, pp. 463–464.

Gulick, V. C., and Baker, V. R., 1993, Fluvial valleys in the heavily cratered terrains of Mars: Evidence for paleoclimatic change? Lunar Planet. Inst. Tech. Rept. 93-03, pp. 12–13.

Haberle, R. M., 1986, The climate of Mars. Sci. Am, 254, no. 5, 54–62.

Haberle, R. M., and Jakosky, B. M., 1990, Sublimation and transport of water from the north residual polar cap on Mars. J. Geophys. Res., 95, 1423–1437.

Haldane, J. B. S., 1929, The origin of life. Rationalist Ann., 148, 3–10.

Hamilton, W., 1972. The Hallet Volcanic Province, Antarctica. U. S. Geol. Survey, Prof. Paper 1534.

Hargraves, R. B., 1976, Precambrian geologic history. Science, 193, 363–371.

Hartmann, W. K., 1973, Martian cratering 4: Mariner 9 initial analysis of cratering chronology. J. Geophys. Res., 78, 4096–4116.

Hess, S. L., Ryan, J. A., Tillman, J. E., Henry, R. M., and Leovy, C. B., 1980, The annual cycle of pressure on Mars measured at Viking 1 and 2. Geophys. Res. Lett., 7, 197–200.

Hodges, C. A., and Moore, H. J., 1979. The subglacial birth of Olympus Mons and its aureoles. J Geophys. Res, 84, 8061–8074.

Hodges, C. A., and Moore, H. J., 1994, Atlas of volcanic landforms on Mars. U. S. Geol. Survey Prof. Paper 1534.

Hoffert, M. I., Calegari, A. J., Hsieh, C. T., and Ziegler, W., 1981, An energy balance climate model for CO_2/H_2O atmospheres. Icarus, 47, 112–129.

Holland, H., 1978, The Chemistry of the Atmosphere and Oceans. Wiley, New York.

Horowitz, N. H., 1986, To Utopia and Back: The Search for Life in the Solar System. W. H. Freeman, New York.

Horton, R. E., 1945, Erosional development of streams and their drainage basins: Hydrophysical approach to quantitative morphology. Geol. Soc. Am. Bull., 56, 275–370.

Howard, A. D., 1978, Origin of the stepped topography of the Martian poles. Icarus, 34, 581–599.

Howard, A. D., Cutts, J. A., and Blasius, K. R., 1982, Stratigraphic relationships within martian polar cap deposits. Icarus, 50, 161–215.

Howard, A. D., Kochel, K. R., and Holt, H. E., 1988, Sapping features of the Colorado Plateau. NASA SP-491.

Huber, R., Stoffers, P., Chminee, J. L., Richnow, H. H., and Stetter, K. O., 1990, Hyperthermophylic archaebacteria within the crater and open-sea plume of erupting MacDonald seamount. Nature, 345, 179–181.

Hunten, D. M., 1993, Atmospheric evolution of the terrestrial planets. Science, 259, 915–920.

Hunten, D. M., Pepin, R. O., and Walker, J. C. G., 1987, Mars fractionation in hydrodynamic escape. Icarus, 69, 532–549.

Hunten, D. M., Pepin, R. O., and Owen, T., 1988, Planetary atmospheres. In Meteorites and the Early Solar System. (Kerridge, J., and Matthews, M., eds). University of Arizona Press, Tucson, AZ, pp. 565–591.

Hunten, D. M., Donahue, T. M., Walker, J.G.C., and Kasting, J. F., 1989, Escape of atmospheres and loss of water. In Origin and Evolution of Planetary and Satellite Atmospheres (Atreya, S. K., et al., eds), University of Arizona Press, Tucson, pp. 386–422.

International Mars Exploration Working Group, 1995, International Strategy for the Exploration of Mars. Planetary and Space Science (in press).

Iverson, R. M., 1985, A constitutive equation for mass-movement behavior. J. Geol., 93, 143–160.

Jagger, J., 1985, Solar-UV Actions on Living Cells. Praeger, New York.

Jagoutz, E., 1991, Chronology of SNC meteorites. Space Science Rev., 56, 13–22.

Jagoutz, E., Baddenhausen, H., Palme, H., Blum, K., Cendales, J., Dreibus, G., Spettel, B., Lorenz, V., and Wänke, H., 1979, The abundances of major, minor and trace elements in the Earth's mantle as derived from primitive ultramafic nodules. Proc. Lunar Planet. Sci. Conf., 10th, pp. 2031–2050.

Jakosky, B. M., 1990, Mars atmospheric D/H: Consistent with polar volatile theory? J. Geophys. Res., 95, 1475–1480.

Jakosky, B. M., and Ahrens, T. J., 1979, The history of an atmosphere of impact origin. Proc. Lunar Planet. Sci. Conf., 10th, pp. 2727–2739.

Jakosky, B. M., and Barker, E. S., 1984, Comparison of groundbased and Viking Orbiter measurements of water vapor: Variability of the seasonal cycle. Icarus, 57, 322–334.

Jakosky, B. M., and Carr, M. H., 1985, Possible precipitation of ice at low latitudes of Mars during periods of high obliquity. Nature, 315, 559–561.

Jakosky, B. M., and Farmer, C. B., 1982, The seasonal and global behavior of water vapor in the Mars atmosphere: Complete global results of the Viking atmospheric water vapor detector experiment. J. Geophys. Res., 87, 2999–3019.

Jakosky, B. M., and Haberle, R. M., 1990, Year-to-year instability of the south polar cap. J. Geophys. Res., 95, 1359–1365.

Jakosky, B. M., and Haberle, R. M., 1992, The seasonal behavior of water on Mars. In Mars (Kieffer, H. H., et al., eds) University of Arizona Press, Tucson, pp. 969–1016.

Jakosky, B. M., Henderson, B. G., and Mellon, M. T., 1993, The Mars water cycle at other epochs: Recent history of the polar caps and layered terrain. Icarus, 102, 286–297.

Jakosky, B. M., Pepin, R. O., Johnson, R. E., and Fox, J . L., 1994, Mars atmospheric loss and isotopic fractionation by solar wind induced sputtering and photochemical escape. Icarus, 111, 271–288.

Jakosky, B. M., Henderson, B. G, and Mellon, M. T., 1995, Chaotic obliquity and the nature of the martian climate. J. Geophys. Res. (in press)

James, P. B., 1985, The Mars hydrologic cycle: Effects of CO_2 mass flux on global water distribution. Icarus, 64, 249–264.

Jarvis, R. S., 1977, Drainage network analysis. Progr. Phys. Geogr., 1, 271–295.

Johanson, L., 1979, The Latitude Dependence of Martian Splosh Craters and its Relationship to Water. NASA Tech. Memo. 80339, 123–124

Johnson, A. M., 1970, Physical Processes in Geology. Freeman, Cooper & Co., San Francisco.

Jones, J. H., 1986, A discussion of isotopic systematics and mineral zoning in the shergottites: Evidence for a 180 m.y. igneous crystallization age. Geochim. Cosmochim. Acta, 50, 969–977.

Jöns, H. -P., 1985 Late sedimentation and late sediments in the northern lowlands of Mars. Lunar Planet. Sci. Conf. XVI, pp. 414–415.

Kahn, R. A., Martin, T. Z., Zurek, R. W., and Lee, S. W., 1992, The martian dust cycle. In Mars (Kieffer, H. H., et al. eds). University of Arizona Press, Tucson, pp. 1017–1053.

Kargel, J. S., and Strom, R. G., 1992, Ancient glaciation on Mars. Geology, 20, 3–7.

Karlsson, H. R., Clayton, R. N., Gibson, E. K., and Mayeda, T. K., 1992, Water in SNC meteorites: Evidence for a martian hydrosphere. Science, 255, 1409–1411.

Kasting, J. F., 1987, Theoretical constraints on oxygen and carbon dioxide concentrations in the Precambrian atmosphere. Precambrian Res., 87, 3091-3098.

Kasting, J. F., 1991, CO_2 condensation and the climate of early Mars. Icarus, 94, 1–13.

Kasting, J. F., and Toon, O. B., 1989, Climate evolution on the terrestrial planets. In Origin and Evolution of Planetary and Satellite Atmospheres. (Atreya, S. K., et al., eds). University of Arizona Press, Tucson, pp. 423–449.

Kieffer, H. H., and Zent, A. P., 1992, Quasi-periodic climate change on Mars. In Mars (Kieffer, H. H., et al., eds), University of Arizona Press, Tucson, pp. 1180–1233.

Kieffer, H. H., Chase, S. C., Martin, T. Z., Miner, E. D., and Palluconi, F. D., 1976, Martian north pole summer temperatures: Dirty water ice. Science, 194, 1341–1344.

Kieffer, H. H., Martin, T. Z., Peterfreund, A. R., Jakosky, B. M., Miner, E. D., and Palluconi, F. D., 1977, Thermal and albedo mapping of Mars during the Viking primary mission. J. Geophys. Res., 82, 4249–4291.

Kieffer, H. H., Jakosky, B. M., Snyder, C. W., and Matthews, M. S., 1992, Mars. University of Arizona Press, Tucson.

Kissel, J., and Kreuger, F. R., 1987, The organic component in the dust from comet Halley as measured by the PUMA mass spectrometer on board VEGA 1. Nature, 326, 755–760.

Klein, H. P., 1978, The Viking biological experiments on Mars. Icarus, 34, 666–674.

Klein, H. P., 1979, The Viking mission and the search for life on Mars. Rev. Geophys., 17, 1655–1662.

Koch, R. P., Zablowski, P., Spreinat, A., and Antranikian, G., 1990, Extremely thermophile amylolytic enzyme from the Archaebacterum *Pyrococcus furiosus.* FEBS Microbiol. Lett., 71, 21–26.

Komar, P. D., 1979, Comparisons of the hydraulics of water flows in martian outflow channels with flows of similar scale on Earth. Icarus, 1980, 37, 156–181.

Komar, P. D., 1980, Modes of sediment transport in channelized water flows with ramifications to the erosion of the martian outflow channels. Icarus, 42, 317–329.

Krasnapolsky, V. A., Bowyer, S., Chakrabarti, S., Gladstone, G. R., and McDonald, J. S., 1994, First measurement of helium on Mars: Implications for the problem of radiogenic gases on the terrestrial planets. Icarus, 109, 337–351.

Kuhn, W. R., and Atreya, S. K., 1979, Ammonia photolysis and the greenhouse effect in the primordial atmosphere of the Earth. Icarus, 37, 207–213.

Kuzmin, R. O., 1983, The Cryolithosphere of Mars. Nauka, Moscow (in Russian).

Kuzmin, R. O., Bobina, N. N., Zabulueva, E. V., and Shaskina, V. P., 1988, Inhomogeneities in the upper levels of the martian cryolithosphere. Lunar Planet. Sci. Conf. XIX, pp. 655–656.

Kuzmin, R. O., Bobina, N. N., Zabulueva, E. V., and Shaskina, V. P., 1989, Martian cryolithosphere: Structure and relative ice content. Int. Geol. Congr., Abs., 2, 245.

Laity, J. E., and Malin, M. C., 1985, Sapping processes and the development of theater-headed valley networks in the Colorado Plateau. Geol. Soc. Am. Bull., 96, 203–217.

Larimer, J. W., 1988, The cosmochemical classification of the elements. In Meteorites and the Early Solar System (Kerridge, J., and M. Matthews, eds), Univversity of Arizona Press, Tucson, pp. 375–389.

Laskar, J., and Robutel, P., 1993, The chaotic obliquity of the planets. Nature, 362, 608–612.

Lee, P., 1993, Briny lakes on early Mars? Terrestrial intracrater playas and martian candidates. Lunar Planet. Inst. Tech. Rept. 93-03, p. 17.

Leighton, R. B., Murray, B. C., Sharp, R. P., Allen, J. D., and Sloan, R. K., 1965, Mariner IV photography of Mars: Initial results, Science, 149, 627–630.

Leighton, R. B., Horowitz, N. H., Murray, B. C., Sharp, R. P., Herriman, A. G., Young, A. T., Smith, B. A., Davies, M. E., and Leovy, C. G., 1969, Mariner 6 and 7 television pictures: Preliminary analysis. Science, 166, 49–67.

Leovy, C. B., and Zurek, R. W., 1979, Thermal tides and martian dust storms: Direct evidence for coupling. J. Geophys. Res., 84, 2956–2968.

Levin, G. V., 1988, A reappraisal of life on Mars. Am. Astron. Soc., 71, 187–207.

Levin, G. V., and Straat, P. A., 1981, A search for nonbiological explanation of the Viking Labeled Release Life Detection Experiment. Icarus, 45, 494–516.

Levine, J. S., Gregory, G. L., Harvey, G. A., Howell, W. E., Borucki, W. J., and Orville, R. E., 1982, Production of nitric oxide by lightning on Venus. Geophys. Res. Lett., 9, 893–896.

Lowell, P., 1895, Mars. Houghton Mifflin, Boston.

Lowell, P., 1906, Mars and its Canals, Macmillan, New York.

Lowell, P., 1908, Mars and the Abode of Life. Macmillan, New York.

Lucchitta, B. K., 1980, Martian outflow channels sculpted by glaciers. Lunar Planet Sci. Conf. XI, pp. 634–636

Lucchitta, B. K., 1981, Mars and Earth: Comparison of cold climate features. Icarus, 45, 264–303.

Lucchitta, B. K., 1982, Ice sculpture in the martian outflow channels. J. Geophys. Res., 87, 9951–9973.

Lucchitta, B. K., 1984, Ice and debris in the fretted terrain, Mars. J. Geophys. Res., 89, B409–B418.

Lucchitta, B. K., Ferguson, H. M., and Summers, C., 1986, Sedimentary deposits in the northern lowland plains, Mars. J. Geophys. Res., 91, E166–E174.

Lucchitta, B. K., McEwen, A. S., Clow, G. D., Geissler, P. E., Singer, R. B., Schultz, R. A., and Squyres, S. W., 1992, The canyon system on Mars. In Mars (Kieffer, H. H., et al., eds) University of Arizona Press, Tucson, pp. 453–492.

Lucchitta, B. K., Isbell, N. K., and Howington-Kraus, A., 1994, Topography of Valles Marineris: Implications for erosional and structural history. J. Geophys. Res., 99, 3783–3798.

Luhmann, J. G., and Kozyra, J. U., 1991, Dayside pickup oxygen ion precipitation at Venus and Mars: Spatial distributions, energy deposition and consequences. J. Geophys. Res., 96, 5457–5467.

Luhmann, J. G., Russel, C. T., Brace, L. H., and Vaisberg, O. L., 1992, The intrinsic magnetic field and solar-wind interactions of Mars. In Mars (Kieffer, H. H., et al., eds), University of Arizona Press, Tucson, pp. 1090–1134.

Madduma Bandara, C. M., 1974, Drainage density and effective precipitation. J. Hydrol., 21, 187–190.

Malin, M. C., 1977, Comparison of volcanic features: Elysium (Mars) and Tibetsi (Earth). Geol. Soc. Amer. Bull, 84, 908–919.

Malin, M. C., 1986, Density of north polar layered terrain deposits: Implications for composition. Geophys. Res. Lett., 13, 444–447.

Mancinelli, R. L., and McKay, C. P., 1988, The evolution of nitrogen cycling. Orign. Life Evol. Biosph., 18, 311–325.

Marov, M. Y., and Petrov, G. I., 1973, Investigations of Mars from the Soviet automatic stations Mars 2 and 3. Icarus, 19, 163–179.

Mars Channel Working Group, 1983, Channels and valleys on Mars. Geol. Soc. Am. Bull., 94, 1035–1054.

Mars Science Working Group, 1991, A Strategy for the Scientific Exploration of Mars. Jet Propulsion Lab., D-8211, Pasadena, CA.

Masursky, H., 1973, An overview of geologic results from Mariner 9. J. Geophys. Res., 78, 4037–4047.

Masursky, H., Boyce, J. V., Dial, A. L., Schaber, G. G., and Strobell, M. E., 1977, Classification and time of formation of Martian channels based on Viking data. J. Geophys. Res., 82, 4016–4037.

Matsui, T., and Abe, Y., 1986, Evolution of an impact induced atmosphere and magma ocean on the accreting Earth. Nature, 319, 303–305.

Matsui, T., and Abe, Y., 1987, Evolutionary tracks of the terrestrial planets. Earth, Moon Planets, 39, 207–214.

McCauley, J. F., Carr, M. H., Cutts, J. A., Hartmann, W. K., Masursky, H., Milton, D. J., Sharp, R. P., and Wilhelms, D. E., 1972, Preliminary Mariner 9 report on the geology of Mars. Icarus, 17, 289–327.

McCauley, J. F., 1978, Geologic map of the Coprates quadrangle of Mars. U. S. Geol. Survey Misc. Inv. Ser. Map I-897.

McElroy, M. B., Kong, T. Y., and Yung, Y. L., 1977, Photochemistry and evolution of Mars' atmosphere: A Viking perspective. J. Geophys. Res., 82, 4379–4388.

McGaw, R. W., and Tice, A. R., 1976, A simple procedure to calculate the volume of water remaining unfrozen in freezing soil. Proc. 2nd Conf on Soil Water Problems in Cold Regions, Edmonton, Alberta, pp. 114–122.

McGill, G. E., 1985, Age and origin of large martian polygons. Lunar Planet. Sci. Conf. XVI, pp. 534–535.

McGill, G. E., 1986, The giant polygons of Utopia, northern martian plains. Geophys. Res. Lett., 13, 705–708.

McKay, C. P., 1991, Urey Prize Lecture: Planetary evolution and the origin of life. Icarus, 91, 93–100.

McKay, C. P., and W. L. Davis, 1991, Duration of liquid water habitats on early Mars. Icarus, 90, 214–221.

McKay, C. P., and Stoker, C. R., 1989, The early environment and its evolution on Mars: Implications for life. Rev. Geophys., 27, 189–214.

McKay, C. P., Clow, G. D., Wharton, R. A., and Squyres, S. S., 1985, Thickness of ice on perennially frozen lakes. Nature, 313, 561–562.

McSween, H. Y., 1985, SNC meteorites: Clues to martian petrologic evolution. Rev. Geophys. 23, 391–416.

McSween, H. J., 1994, What have we learned about Mars from SNC meteorites. Meteoritics, 29, 757–779.

Mellon, M. T., and Jakosky, B. M., 1993, Geographic variations in the thermal and diffusive stability of ground ice on Mars. J. Geophys. Res., 98, 3345–3364.

Melosh, H. J., 1984, Impact ejection, spallation and the origin of meteorites. Icarus, 59, 234–260.

Melosh, H. J., and Vickery, A. M., 1989, Impact erosion of the primordial atmosphere of Mars. Nature, 338, 487–489.

Michaux, C. M., and Newburn, R. L., 1972, Mars Scientific Model. Jet Propulsion Laboratory, Document 606-1.

Miller, S. L., 1953, A production of amino acids under possible primitive Earth conditions. Science, 117, 528–529.

Miller, S. L., 1955, Production of some organic compounds under possible primitive Earth conditions, J. Am. Chem. Soc., 77, 2352–2361.

Miller, S. L., 1957, The mechanism of synthesis of amino acids by electric discharges. Biochim. Biophys. Acta, 23, 480–489.

Miller, S. L., and Smythe, W. D., 1970, Carbon dioxide clathrate in the martian ice cap. Science, 170, 531–533.

Mittelfehldt, D. W., 1994, ALHA84001, a cumulate orthopyroxenite member of the martian meteorite clan. Meteoritics, 29, 2114–221.

Mouginis-Mark, P. J., 1979, Marian fluidized crater morphology: Variations with crater size, latitude, altitude, and target material. J. Geophys. Res., 84, 8011–8022.

Mouginis-Mark, P. J., 1985, Volcano-ground ice interactions in Elysium Planitia. Icarus, 64, 265–284.

Mouginis-Mark, P. J., and Cloutis, E. A., 1983, Ejecta areas of impact craters on the martian ridged plains. Lunar Planet. Sci. Conf. XIV, pp. 532–533.

Mouginis-Mark, P. J., Wilson, L., and Zimbelman, J. R., 1988, Polygenic eruptions on Alba Patera, Mars: Evidence of channel erosion on pyroclastic flows. Bull. Volcanol., 50, 361–379.

Murray, B. M., Soderblom, L. A., Cutts, J. A., Sharp, R. P., Milton, D. J., and Leighton, R. B., 1972, Geologic framework of the south polar region of Mars. Icarus, 17, 328–345.

Murthy, V. R., 1991, Early differentiation of the Earth and problem of mantle siderophile elements: A new approach. Science, 253, 303–306.

Mutch, T. A., Arvidson, R. E., Head, J. W., Jones, K. L., and Saunders, R. S., 1976, The Geology of Mars. Princeton University Press, Princeton, NJ.

Mutch, T. A., Arvidson, R. E., Binder, A. B., Guiness, E. A., and Morris, E. M., 1977, The geology of the Viking Lander 2 site. J. Geophys. Res., 82, 4452–4467.

National Research Council, 1992, Biological Contamination of Mars, National Academy Press, Washington, D. C.

Nedell, S. S., Squyres, S. W., and Andersen, D. W., 1987, Origin and evolution of the layered deposits in the Valles Marineris, Mars. Icarus, 70, p. 409–441.

Neukum, G., and Wise, D. U., 1976, Mars: A standard crater curve and possible new time scale. Science, 194, 1381–1387.

Newman, M. J., and Rood, R. T., 1977, Implications of solar evolution for Earth's early atmosphere. Science, 198, 1035–1037.

Newsom, H. E., 1980, Hydrothermal alteration of impact melt sheets with implications for Mars. Icarus, 44, 207–216.

Newsom, H. E., 1990, Accretion and core formation in the Earth: Evidence from siderophile elements. In Origin of the Earth (H. E., Newsom and J. H. Jones, eds), Oxford University Press, New York, pp. 273–288.

Nummedal, D., 1978, The Role of Liquefaction in Channel Development on Mars. NASA Tech. Memo 79729, pp. 257–259.

Nyquist, L. E., Bansal, B. M., Wiesman, H., and Shih, C.-Y., 1995, "Martians" young and old: Zagami and ALH84001. Lunar Planet Sci. Conf. XXXI, pp. 1065–1066.

Oparin, A. I., 1957, The Origin of Life on the Earth. Academic Press, New York.

Owen, T., 1992, Composition and early history of the atmosphere. In Mars (Kieffer, H. H., et al. eds), University of Arizona Press, Tucson, pp. 818–834.

Owen, T., Niemann, K., Rushneck, D. R., Biller, J. E., Howarth, D. W., and LaFleur, A. L., 1977, The composition of the atmosphere at the surface of Mars. J. Geophys. Res., 82, 4635–4639.

Owen, T., Cess, R. D., and Rammanathan, V., 1979, Early Earth: An enhanced carbon dioxide greenhouse to compensate for reduced solar luminosity. Nature, 277, 640–642.

Owen, T., Maillard, J. P., DeBergh, C., and Lutz, B. L., 1988, Deuterium on Mars: The abundance of HDO and the value of D/H. Science, 240, 1767–1770.

Owen, T., Bar-Nun, A., and Kleinfeld, I., 1992, Possible cometary origin of heavy noble gases in the atmospheres of Venus, Earth and Mars. Nature, 358, 43–46.

Pace, N. R., 1991, Origin of life—facing up to the physical setting. Cell, 65, 531–533.

Paige, D. A., 1992, The thermal stability of near-surface ground ice on Mars. Nature, 356, p. 43–45.

Palluconi, F. D., 1979, Mars: The thermal inertia of the surface from $-60°$ to $+60°$. Bull. Am. Astron. Soc., 11, 575.

Palluconi, F. D., and Kieffer, H. H., 1981, Thermal inertia mapping of Mars for 60°S to 60°N. Icarus, 45, 415–426.

Palmer, A. N., 1990, Groundwater Processes in Karst Terranes. Geol. Soc. Am. Spec. Paper 252, pp. 177–209.

Parker, T. J., Saunders, R. S., and Schneeberger, D. M., 1989, Transitional morphology in the west Deuteronilus Mensae region of Mars: Implications for modification of the lowland/upland boundary. Icarus, 82, 111–145.

Parker, T. J., Gorcine, D. S., Saunders, R. S., Pieri, D. C., and Schneeberger, D. M., 1993, Coastal geomorphology of the martian northern plains. J. Geophys. Res., 98, 11,061–11,078.

Pechmann, J. C., 1980, The origin of polygonal troughs on the northern plains of Mars. Icarus, 42, 185–210.

Pepin, R. O., 1985, Evidence of martian origins. Nature, 317, 473–475.

Pepin, R. O., 1991, Evolution of atmospheric volatiles. Icarus, 92, 2–79.

Pepin, R. O., 1994, Evolution of the martian atmosphere. Icarus, 111, 289–304.

Pieri, D. C., 1976, Martian channels: Distribution of small channels on the martian surface. Icarus, 27, 25–50.

Pieri, D. C., 1980a, Martian valleys: Morphology, distribution, age and origin. Science, 895–897.

Pieri, D. C., 1980b, Geomorphology of martian valleys. NASA TM-81979, 4–160.

Plaut, J. J., Kahn, R., Guiness, E. A., and Arvidson, R.E., 1988, Accumulation of sedimentary debris in the south polar region of Mars and implications for climate history. Icarus, 75, 357–377.

Pollack, J. B., 1979, Climate change on the terrestrial planets. Icarus, 37, 479–553.

Pollack, J. B., and Black, D. C., 1979, Implications of the gas compositional measurements of Pioneer Venus for the origin of planetary atmospheres. Science, 205, 56–59.

Pollack, J. B., and Yung, Y. L., 1980, Origin and evolution of planetary atmospheres. Ann. Rev. Earth Planet. Sci., 8, 425–487.

Pollack, J. B., Colburn, D. S., Flaser, M., Kahn, R., Carlston, C. E., and Pidek, D., 1979, Properties and effects of dust particles suspended in the martian atmosphere. J. Geophys. Res., 84, 2929–2945.

Pollack, J. B., Kasting, J. F., Richardson, S. M., and

Poliakoff, K., 1987, The case for a warm, wet climate on early Mars. Icarus, 71, 203–224.

Pollack, J. B., Roush, T., Witteborn, F., Bregman, J., Wooden, D., Stroker, C., Toon, O. B., Rank, D., Dalton, B., and Freedmann, R., 1990, Thermal emission spectra of Mars (5.4–10.5 μm): Evidence for sulfates, carbonates and hydrates. J. Geophys. Res., 14595–14627.

Postawko, S. E., and Kuhn, W. R., 1986, Effect of greenhouse gases (CO_2, H_2O, SO_2) on martian paleoclimates. J. Geophys. Res., 91, D431–D438.

Reimers, C. E., and P. D. Komar, 1979, Evidence for explosive volcanic density currents on certain martian volcanoes. Icarus, 39, 88–110, 1979.

Ringwood, A. E., 1977, Composition of the core and implications for origin of the Earth. Geochem. J., 11, 111–135.

Ringwood, A. E., 1979, Origin of the Earth and the Moon. Springer-Verlag, New York.

Robinson, M. S., and Tanaka, K. L., 1990, Magnitude of a catastrophic flood event at Kasei Valles, Mars. Geology, 18, 902–905.

Rossbacher, L. A., and Judson, S., 1981, Ground ice on Mars: Inventory, distribution and resulting landforms. Icarus, 45, 39–59.

Safranov, V. S., 1972, Evolution of the Protoplanetary Cloud and Fomation of the Earth and the Planets. NASA TT F-677.

Sagan, C., and Mullen, 1972, Earth and Mars: Evolution of atmospheres and surface temperatures. Science, 177, 52–56.

Sagan, C., Toon, O. B., and Gierasch, P. J., 1973, Climate change on Mars. Science, 181, 1045–1049.

Sasaki, S., and Nakazawa, K, 1988, Origin of isotopic fractionation of terrestrial Xe: Hydrodynamic fractionation during escape of primordial H_2-He atmosphere. Earth Planet. Sci. Lett., 89, 323–334.

Saunders, I., and Young, A., 1983, Rates of surface processes on slopes, slope retreat and denudation. Earth Surface Processes Landforms, 8, 473–301.

Scambos, T. A., and Jakosky, B. M., 1990, An ougassing release factor for non-radiogenic volatiles on Mars. J. Geophys. Res., 95, 14779–14787.

Schaefer, M. W., 1990, Geochemical evolution of the northern plains of Mars: Early hydrosphere, carbonate development, and present morphology. J. Geophys. Res., 95, 14291–14300.

Schaefer, M. W., 1993, Volcanic recycling of carbonates on Mars. Geophys. Res. Lett., 20, 827–830.

Schidlowski, M., 1987, Application of stable carbon isotopes to early biochemical evolution on Earth, Ann. Rev. Earth Planet. Sci., 15, 47–72.

Schopf, J. W., and Walter, M. R., 1983, Archean microfossils: New evidence of ancient microbes. In The Earth's Earliest Biosphere (Schopf, J. W., ed), Princeton University Press, Princeton, NJ, pp. 214–239.

Schopf, J. W., Hayes, J. M., and Walter, M. R., 1983, Evolution of Earth's earliest ecosystems: Recent progress and unsolved problems. In The Earth's Earliest Biosphere (Schopf, J. W., ed), Princeton University Press, Princeton, NJ, pp. 361–384.

Schonfeld, E., 1977, Martian volcanism. Lunar Planet. Sci. Conf. VIII, pp. 843–845.

Schubert, G., and Spohn, T., 1990, Thermal history of Mars and the sulfur content of its core. J. Geophys. Res., 95, 14095–14104.

Schubert, G., Solomon, S. C., Turcotte, D. L., Drake, M. J., and Sleep, N. H., 1992, Origin and thermal evolution of Mars. In Mars (Kieffer, H. H., et al., eds), University of Arizona Press, Tucson, pp. 147–183.

Schumm, S. A., 1974, Structural origin of large martian channels. Icarus, 22, p. 371–384.

Schultz, P. H., and Britt, D., 1986, Martian gradation history. Lunar Planet. Sci. Conf. XVII, pp. 775–776.

Schultz, P. H., and Gault, D. E., 1979, Atmospheric effects on Martian ejecta emplacement. J. Geophys. Res., 84, 7669–7687.

Schultz, P. H., and Gault, D. E., 1984, On the formation of contiguous ramparts around martian impact craters. Lunar Planet. Sci. Conf. XV, pp. 732–733.

Schultz, P. H., and Lutz, A. B., 1988, Polar wandering on Mars. Icarus, 73, 91–141.

Sclater, J. G., Jaupart, C., and Galson, D., 1980, The heat flow through oceanic and continental crust and the heat loss of the earth. Rev. Geophys. Space Phys., 18, 269–311.

Scott, D. H., and Chapman, M. G., 1991, Mars Elysium basin: Geologic/volumetric analyses of a young lake and exobiologic implications. Proc. Lunar Planet. Sci., 21, 669–677.

Scott, D. H., and Dohm, J. M., 1992, Mars highland channels: an age reassessment. Lunar Planet. Sci. Conf. XXIII, pp. 1251–1252.

Scott, D. H., and Tanaka, K. L., 1986, Geologic map of the western equatorial region of Mars. U.S. Geological Survey Misc Inv. Map I-1802-A.

Scott, D. H., Chapman, M. G., Rice, J. W., and Dohm, J. M., 1992, New evidence of lacustrine basins on Mars: Amazonis and Utopia Planitia. Proc. Lunar Planet. Sci., 22, 53–62.

Scott, D. H., Dohm, J. M., and Rice, J. W., 1995, Map of Mars showing channels and possible paleolakes. U.S. Geol. Survey, Misc. Inv. Map I-2461.

Sears, D. W. G., and Dodd, R. T., 1988, Overview and classification of meteorites. In Meteorites and the Early Solar System (Kerridge, J., and and Matthews, M., eds) University of Arizona Press, Tucson, pp. 3–31.

Sellin R. H. J., 1969, Flow in Channels. St, Martins Press, London.

Sellers, W. D., 1965, Physical climatology, University of Chicago Press, Chicago.

Sharp, R. P., 1973, Mars: Fretted and chaotic terrains. J. Geophys. Res., 78, 4222–4230

Sharp, R. P., and Malin, M. C., 1975, Channels on Mars. Geol. Soc. Am. Bull, 86, 593–609.

Shih, C. -Y., Nyquist, L. E., Bogard, D. D., McKay, G. A., Wooden, J. L., Bansei, B. L., and Wiesman, H., 1982, Chronology and petrogenesis of young achondries, Shergotty, Zagami and ALHA 77005: Late magmatism on the geologically active planet: Geochim. Cosmochim. Acta, 46, 2323–2344.

Shoji, H., and Higashi, A., 1978, A deformation mechanism map of ice. J. Glaciol. 85, 419–427.

Shreve, R. L., 1966, Statistical law of stream numbers. J. Geol., 74, 17–37.

Shreve, R. L., 1967, Infinite topologically random channel networks. J. Geol., 75, 178–186.

Sleep, N. H., Zahnle, K. J., Kasting, J. F., and Morowits, H. J., 1989, Annihilation of ecosystems by large asteroid impacts on the early Earth. Nature, 342, 139–142.

Slipher, E. C., 1964, A Photographic Study of the Brighter Planets. Lowell Observatory, Flagstaff, AZ.

Sloan, C. E., Zenone, C., and Mayo, I. R., 1976, Icings along the Trans-Alaska pipeline route. U.S. Geological Survey Prof. Paper 979.

Smart, J. S., 1972, Quantitative characterization of channel network structure. Water Resources Res., 8, 1487–1496.

Smith, D. E., and Zuber, M. T., 1994, The topography of Mars: A re-evaluation of current data. Lunar Planet. Sci. Conf. XXV, pp. 1289–1290.

Smoluchowski, R., 1968, Mars: Retention of ice. Science, 159, 1348–1350.

Snyder, C. W., and Moroz, V. I., 1992, Spacecraft exploration. In Mars (Kieffer, H. H. et al. eds), University of Arizona Press, Tucson, pp. 71–119.

Soderblom, L. A., Kriedler, T. J., and Masursky, H., 1973, Latitudinal distribution of debris mantle on the martian surface. J. Geophys. Res., 78, 4117–4122.

Soderblom, L. A., and Wenner, D. B., 1978, Possible fossil liquid-ice interfaces in the Martian crust. Icarus, 34, 622–637.

Solomon, S. C., 1979, Formation, history and energetics of cores in terrestrial planets. Phys. Earth Planet. Int., 19, 168–182.

Spencer, J. R., and Fanale, F. P., 1990, New models for the origin of Vallis Marineris closed depressions. J. Geophys. Res., 95, 14301–14313.

Squyres, S. W., 1978, Mars fretted terrains, flow of erosional debris. Icarus, 34, 600–613.

Squyres, S. W., 1979, The distribution of lobate debris aprons and similar flows on Mars. J. Geophys. Res., 84, 8087–8096.

Squyres, S. W., 1989a, Early Mars: Wet and warm, or just wet. Lunar Planet Sci. Conf. XX, pp. 1044–1045.

Squyres, S. W., 1989b, Urey prize lecture: Water on Mars. Icarus, 79, 229–288.

Squyres, S. W., and Carr, M. H., 1986, Geomorphic evidence for the distribution of ground ice on Mars. Science, 231, 249–252.

Squyres, S. W., and Kasting, J. F., 1994, Early Mars: How warm and how wet? Science, 265, 744–748.

Squyres, S. W., Wilhelms, D. E., and Moosman, A. C., 1987, Large-scale volcano ice interactions on Mars. Icarus, 70, 385–408.

Squyres, S. W., Clifford, S. M., Kuzmin, R. O., Zimbelman, J. R., and Costard, F. M., 1992, Ice in the martian regolith. In Mars (Kieffer, H. H., et al., eds), University of Arizona Press, Tucson, pp. 523–554.

Stevenson, D. J., Spohn, T., and Schubert, G., 1983, Magnetism and thermal evolution of the terrestrial planets. Icarus, 466–489.

Strahler, A. N., 1958, Dimensional analysis applied to fluvially eroded landforms. Geol. Soc. Am. Bull., 69, 279–300.

Strahler, A. N., 1964, Quantitative geomorphology of drainage basins and channel networks. In Handbook of Applied Hydrology (V. T., Chow, ed.), McGraw Hill, New York, pp. 4–39 to 4–76.

Strom, R. G., Croft, S. K., and Barlow, N. D., 1992, The martian impact createring record. In Mars (Kieffer, H. H., et al., eds), University of Arizona Press, Tucson, pp. 383–423

Summers, D. P., and Chang, S., 1993, Prebiotic ammonia from reduction of nitrite by iron (II) on the early Earth. Nature, 365, 630–633.

Tanaka, K. L., 1986, The stratigraphy of Mars. Proc. Lunar Planet. Sci. Conf., 17th, J. Geophys. Res., 91, E139–E158.

Tanaka, K. L., Isbell, N. K., Scott, D. H., Greeley, R., and Guest, J. E., 1988, The resurfacing history of Mars: A synthesis of digitized, Viking-based geology. Proc. Lunar Planet. Sci. Conf., 18th, pp. 665–678.

Tanaka, K. L., Scott, D. H., and Greeley, R., 1992, Global stratigraphy. In Mars (Kieffer, H. H., et al. eds), University of Arizona Press, Tucson, pp. 345–382.

Taylor, R. S., 1988, Planetary compositions. In Meteorites and the Solar System (J. Kerridge and M. Matthews, eds), University of Arizona Press, Tucson, pp. 512–534.

Terzaghi, K., 1943, Theoretical Soil Mechanics. Wiley, New York.

Thomas, P., Squyres, S., Herkenhoff, K., Howard, A., and Murray, B., 1992, Polar deposits of Mars. In Mars (Kieffer, H. H., et al., eds), University of Arizona Press, Tucson, pp. 767–795.

Toksoz, M. N., Dainty, A. M., Solomon, S. C., and Anderson, K. R., 1974, Structure of the Moon. Rev. Geophys., 12, 539–567.

Tolstikhin, N. I., and Tolstikhin, O. N., 1974, Groundwater and surface water in the permafrost region (English translation). Canada Inland Waters Directorate, Tech. Bull. 97, 25 pp.

Toon, O. B., Pollack, J. B., Ward, W., Burns, J. A., and Bilski, K., 1980, The astronomical theory of climate change on Mars. Icarus, 44, 552–607.

Touma, J., and Wisdom, J., 1993, The chaotic obliquity of Mars. Science, 259, 1294–1296.

Treiman, A. H., Drake, M. J., Janssens, M., Wolf, R., and Ebihara, M., 1986, Core formation in the Earth and Shergottite Paren Body (SPB): Chemical evidence from basalts. Geochim. Cosmochim. Acta, 50, 1071–1091.

Wächtershäuser, G., 1988, Before enzymes and templates: Theory of surface metabolism. Microbiol. Rev., 52, 452–484.

Wahrhaftig, C., and Cox, A., 1959, Rock glaciers in the Alaska range. Geo. Soc. Am. Bull., 70, 383–436.

Walker, J. G. C., 1977, Evolution of the Atmosphere. Macmillan, New York.

Walker, J. G. C., 1985, Carbon dioxide on the early Earth. Origins Life, 16, 117–127.

Wallace, D., and Sagan, C., 1979, Evaporation of ice in planetary atmospheres: Ice-covered rivers on Mars. Icarus, 39, 385–400.

Walter, F. M., 1983, Archean stromatolites: Evidence of the Earth's earliest benthos. Earth's Earliest Biosphere (Schopf, J. W., ed.), Princeton University Press, Princeton, NJ, pp. 187–213.

Walter, F. M., Brown, A., Mathieu, R. D., Myers, P. C., and Vrba, F. J., 1988, X-ray sources in regions of star formation . III. Naked T- Tauri stars associated with the Taurus-Auriga complex. Astron. J., 96, 297–325.

Wänke, H., 1981, Constitution of terrestrial planets. Phil. Trans. R. Soc. Lond, A 303, 287–303

Wänke, H., and Dreibus, G., 1988, Chemical composition and accretion history of terrestrial planets. Phil. Trans. R. Soc. Lon, A 325, 545–557.

Ward, W. R., 1973, Large scale variations in the obliquity of Mars. Science, 181, 260–262.

Ward, W. R., 1992, Long term orbital and spin dynamics of Mars. In Mars (Kieffer, H. H., et al., eds), University of Arizona Press, Tucson, pp. 298–320.

Ward, W. R., Burns, J. A., and Toon, O. B., 1979, Past obliquity oscillations of Mars: The role of the Tharsios uplift. J. Geophys. Res., 84, 243–259.

Washburn, A. L., 1967, Instrumental observations of mass wasting in the Mesters Veg district, Northeast Greenland. Medd. Grønland, 166, 318 pp.

Washburn, A. L., 1973, Periglacial Processes and Environments. St. Martins, New York.

Washburn, A. L., 1980, Geocryology. Wiley, New York.

Watson, L. L., Hutcheon, I. D., Epstein, S., and Stolper, E. M., 1994, Water on Mars: Clues from deuterium/hydrogen and water contents of hydrous phases in SNC meteorites. Science, 265, 86–90.

Weidenschilling S, J., 1984, Evolution of grains in a turbulent solar nebula. Icarus, 60, 553–567.

Wetherill, G. W., 1990, Formation of the Earth. Annual Rev. Earth Planet. Sci., 18, 205–256.

Wetherill, G. W., and Stewart, G. R., 1989, Accumulation of a swarm of small planetesimals. Icarus, 77, 330–357.

Wiens, R. C., and Pepin, R. O., 1988, Laboratory shock emplacement of noble gases and carbon dioxide into basalt and implications for trapped gases in shergottite EETA 79001. Geochim. Cosmochim. Acta, 52, 295–307.

Wiens, R. C., Becker, R. H., and Pepin, R. O., 1986, The case for martian origin of the shergottites. II. Trapped and indigenous gas components in EETA 79001 glass. Earth Planet. Sci. Lett. 77, 149–158.

Wilhelms, D. E., 1987, The geologic history of the Moon. U.S. Geological Survey, Prof. Paper 1348.

Williams, P. J., and Smith, M. W., 1989, The Frozen Earth. Cambridge University Press, Cambridge.

Wilhelms, D. E., and Squyres, S. W., 1984, The martian hemispheric dichotomy may be due to a giant impact. Nature, 309, 138–140.

Wilson, L., and Head, J. W., 1994, Mars: Review and analysis of volcanic eruption theory and relationships to observed landforms. Rev. Geophys., 32, 221–263.

Wilson, L, and Mouginis-Mark, P. J., 1987, Volcanic input to the atmosphere from Alba Patera on Mars. Nature, 330, 354–357.

Woese, C. R, 1987, Bacterial evolution. Microbiol. Rev., 51, 221–271.

Woese, C. R., 1990, Toward a natural system of organisms. Proc. Natl. Acad. Sci. USA, 87, 4576–4579.

Wood, C. A., and Ashwal, L. D., 1981, SNC meteorites: Igneous rocks from Mars. Proc. Lunar Planet. Sci. Conf., 12B, 1359–1375.

Wood, J. A., 1979, The Solar System. Prentice Hall, Englewood Cliffs, NJ.

Wu, S. S. C., 1978, Mars synthetic topographic mapping. Icarus, 33, 417–440.

Yayonos, A., 1986, Evolutional and ecological implications of the properties of deep-sea barophilic bacteria. Proc. Natl. Acad. Sci. USA, 83, 9542–9546.

Young, D. T., Freidhoff, C. B., Jakosky, B. M., Owen, T. C., and Kasting, J. F., 1994, Understanding the Evolution and Behavior of the Martian Atmosphere and Climate. Jet Propulsion Lab. Tech. Rept. D-12017, pp. 168–169.

Yung, Y. L., and Pinto, J. P., 1978, Primitive atmosphere and implications for the formation of channels on Mars. Nature, 273, 730–732.

Yung, Y. L., Wen, J., Pinto, J. P., Allen, M., Pierce, K. K., and Paulsen, S., 1988, HDO in the martian atmosphere: Implications for the abundance of crustal water. Icarus, 76, 146–159.

Zahnle, K. J., and Kasting, J. F., 1986, Mass fractionation during transonic hydrodynamic escape and implications for loss of water from Venus and Mars. Icarus, 68, 462–480.

Zahnle, K. J., and Walker, J. C. G., 1982, The evolution of solar ultraviolet luminosity. Rev. Geophys. Space Phys., 20, 280–292.

Zahnle, K. J., Kasting, J. F., and Pollack, J. B., 1988, Evolution of a steam atmosphere during Earth's accretion. Icarus, 74, 62–97.

Zahnle, K. J., Kasting, J. F., and Pollack, J. B., 1990, Mass fraction of noble gases in diffusion-limited hydrodynamic hydrodynamic escape. Icarus, 84, 502–527.

Zent, A. P., Fanale, F. P., Salvail, J. R., and Postawko, S. E., 1986, Distribution and state of H_2O in the high-latitude shallow subsurface of Mars. Icarus, 67, 19–36.

Zent, A. P., Quinn, R. C., and Jakosky, B. M., 1994, Fractionation of nitrogen isotopes on Mars: The role of the regolith as a buffer. Icarus, 112, 537–540.

Zhang, M. H. G., Luhmann, J. G., Bougher, S. W., and Nagy, A. F., 1993, The ancient oxygen exosphere of Mars: Implications for atmospheric evolution. J. Geophys. Res., 98, 10915–10923.

Zimbelman, J. R., Clifford, S. M., and Williams, S. H., 1988, Terrain softening revisited: Photogeological considerations. Lunar Planet. Sci. Conf. XIX, pp. 1321–1322.

Zurek, R. W., 1992, Comparative aspects of the climate of Mars: An introduction to the current atmosphere. In Mars (Kieffer, H. H., et al., eds), University of Arizona Press, Tucson, pp. 799–817

Zurek, R. W., Barnes, J. R., Haberle, R. M., Pollack, J. B., Tillman, J. E., and Leovy, C. B., 1992, Dynamics of the atmosphere of Mars. In Mars (Kieffer, H. H., et al., eds), University of Arizona Press, Tucson, pp. 835–933.

INDEX